Green Nanomaterials

From bioinspired synthesis to sustainable manufacturing of inorganic nanomaterials

Green Nanomaterials

From bioinspired synthesis to sustainable manufacturing of inorganic nanomaterials

Siddharth V Patwardhan
Green Nanomaterials Research Group, Department of Chemical and Biological Engineering, The University of Sheffield, UK

Sarah S Staniland
Bio-Nanomagnetic Research Group, Department of Chemistry, The University of Sheffield, UK

IOP Publishing, Bristol, UK

© Siddharth V Patwardhan and Sarah S Staniland 2020

All rights reserved. No part of this publication may be reproduced, stored in a retrieval system or transmitted in any form or by any means, electronic, mechanical, photocopying, recording or otherwise, without the prior permission of the publisher, or as expressly permitted by law or under terms agreed with the appropriate rights organization. Multiple copying is permitted in accordance with the terms of licences issued by the Copyright Licensing Agency, the Copyright Clearance Centre and other reproduction rights organizations.

Certain images in this publication have been obtained by the authors from Wikimedia websites, where they were made available under a Creative Commons licence or stated to be in the public domain. Please see individual figure captions in this publication for details. To the extent that the law allows, IOP Publishing disclaim any liability that any person may suffer as a result of accessing, using or forwarding the images. Any reuse rights should be checked and permission should be sought if necessary from Wikimedia and/or the copyright owner (as appropriate) before using or forwarding the images.

Permission to make use of IOP Publishing content other than as set out above may be sought at permissions@ioppublishing.org.

Siddharth V Patwardhan and Sarah S Staniland have asserted their right to be identified as the authors of this work in accordance with sections 77 and 78 of the Copyright, Designs and Patents Act 1988.

ISBN   978-0-7503-1221-9 (ebook)
ISBN   978-0-7503-1222-6 (print)
ISBN   978-0-7503-1836-5 (myPrint)
ISBN   978-0-7503-1223-3 (mobi)

DOI   10.1088/978-0-7503-1221-9

Version: 20191201

IOP ebooks

British Library Cataloguing-in-Publication Data: A catalogue record for this book is available from the British Library.

Published by IOP Publishing, wholly owned by The Institute of Physics, London

IOP Publishing, Temple Circus, Temple Way, Bristol, BS1 6HG, UK

US Office: IOP Publishing, Inc., 190 North Independence Mall West, Suite 601, Philadelphia, PA 19106, USA

Contents

Preface	x
Acknowledgements	xii
Author biographies	xiii

Section I Green chemistry principles

1	**Green chemistry and engineering**	**1-1**
1.1	Principles of green chemistry and engineering	1-1
	1.1.1 Overview	1-1
	1.1.2 Drivers for green approaches	1-2
	1.1.3 Estimating environmental impact	1-4
1.2	Ways to improve sustainability	1-6
1.3	Green chemistry and nanomaterials	1-7
	References	1-8

Section II Nanomaterials

2	**Nanomaterials: what are they and why do we want them?**	**2-1**
2.1	Fundamentals of the nanoscale	2-1
2.2	Tangible and historical examples of nanomaterials	2-4
2.3	Special properties offered by the nanoscale	2-6
	2.3.1 Optical: surface plasmon resonance	2-7
	2.3.2 Optical: quantum dots fluorescence	2-9
	2.3.3 Electron spin and nanomagnetism	2-11
2.4	Applications	2-14
	2.4.1 Nanomedicine	2-15
	2.4.2 Nanodevice technologies	2-25
	2.4.3 Consumer products	2-30
2.5	Nanomaterial biocompatibility and toxicity	2-31
2.6	Summary: key lessons from nanomaterials, nanoproperties and applications	2-36
	References	2-38

3 Characterisation of nanomaterials — 3-1

3.1 Introduction — 3-1
3.2 Microscopy — 3-2
    3.2.1 Optical microscopy — 3-2
    3.2.2 Electron microscopy — 3-3
    3.2.3 Scanning electron microscopy — 3-4
    3.2.4 Transmission electron microscopy — 3-5
    3.2.5 Atomic force microscopy — 3-7
3.3 Spectroscopy applied to nanomaterials — 3-8
    3.3.1 Mass spectrometry — 3-8
    3.3.2 Infra-red spectroscopy — 3-10
    3.3.3 X-ray photoelectron spectroscopy — 3-13
3.4 Diffraction and scattering techniques — 3-14
    3.4.1 X-ray diffraction — 3-14
    3.4.2 Dynamic light scattering — 3-16
    3.4.3 Small angle scattering — 3-16
3.5 Porosimetry — 3-17
3.6 Summary: key lessons for characterisation of nanomaterials — 3-20
    References — 3-21

4 Conventional methods to prepare nanomaterials — 4-1

4.1 Top-down and bottom-up methods — 4-1
4.2 Top-down methods — 4-3
4.3 Bottom-up methods — 4-4
4.4 Nucleation and growth theory — 4-5
    4.4.1 Homogeneous nucleation — 4-6
    4.4.2 Heterogeneous nucleation — 4-7
    4.4.3 Growth — 4-8
4.5 Conventional bottom-up methods — 4-11
    4.5.1 Vapour-phase method — 4-11
    4.5.2 Solution processing — 4-11
    4.5.3 Spray conversion — 4-12
    4.5.4 Sol–gel method — 4-12
4.6 Emerging bottom-up methods — 4-12
    4.6.1 Principles and overview — 4-12
    4.6.2 Soft lithography — 4-14
    4.6.3 Dip-pen nanolithography — 4-16

	4.6.4 Layer-by-layer self-assembly	4-16
	4.6.5 Solution synthesis of nanoparticles	4-18
	4.6.6 Templated synthesis	4-19
4.7	Summary: key lessons about conventional routes to nanomaterials	4-20
	References	4-20

Section III From biominerals to green nanomaterials

5 Green chemistry for nanomaterials 5-1

5.1	Sustainability of nanomaterials production	5-1
5.2	Reasons behind unsustainability	5-2
5.3	Evaluation of sustainability for selected methods	5-3
	5.3.1 E-factors for solution methods	5-3
	5.3.2 How green is soft lithography?	5-4
	5.3.3 Templated synthesis: surely sustainable?	5-5
5.4	Adopting green chemistry for nanomaterials	5-5
5.5	Biological and biochemical terminology and methods	5-6
	5.5.1 Molecular biology component	5-6
	5.5.2 Molecular biological techniques	5-11
5.6	Summary: key lessons about sustainability nanomaterials production	5-12
	References	5-12

6 Biomineralisation: how Nature makes nanomaterials 6-1

6.1	Introduction	6-1
6.2	Properties and function of biomineral types	6-4
	6.2.1 Bio-calcium phosphate (hydroxyapatite): mechanical/structural support, motion, cutting/grinding	6-4
	6.2.2 Bio-calcium carbonate: protection, sensor, buoyancy	6-6
	6.2.3 Bio-silica: mechanical support, transport and protection	6-7
	6.2.4 Bio-magnetite: sensing, cutting/grinding, iron storage	6-8
6.3	Mineral formation controlling strategies in biomineralisation	6-10
	6.3.1 The universal biomineralisation process	6-11
6.4	Roles and types of organic biological components required for biomineralisation	6-12
	6.4.1 Roles of organic biological components	6-12
	6.4.2 Types of organic biological components	6-14
6.5	Summary: key lessons from biomineralisation for the green synthesis of nanomaterials	6-19
	References	6-20

7 Bioinspired 'green' synthesis of nanomaterials   7-1

7.1	From biological to bioinspired synthesis	7-1
7.2	Mechanistic understanding	7-3
	7.2.1 Biomineralising biomolecules	7-3
	7.2.2 Abiotic peptides and proteins from biopanning	7-5
7.3	An illustration of exploiting the knowledge of nano–bio interactions	7-9
7.4	Additives as the mimics of biomineral forming biomolecules	7-13
	7.4.1 The need for additives	7-13
	7.4.2 The design of additives and custom synthesis	7-14
7.5	Compartmentalisation, templating and patterning	7-18
	7.5.1 Confinement in a simple protein template	7-19
	7.5.2 Confinement in modified cage protein templates	7-22
	7.5.3 Biomimetic compartmentalisation	7-23
	7.5.4 Localisation and patterning on surfaces	7-25
7.6	Scalability of bioinspired syntheses	7-27
7.7	Summary: key lessons about the journey towards bioinspired synthesis	7-28
	References	7-29

Section IV Case studies

8 Case study 1: magnetite nanoparticles   8-1

8.1	Magnetite biomineralisation in magnetotactic bacteria	8-1
8.2	Magnetosome use in applications: advantages and drawbacks	8-3
8.3	Biomolecules and components controlling magnetosome formation	8-4
	8.3.1 Magnetosome biomineralisation protein discovery	8-4
	8.3.2 Bio-components for each step of biomineralisation	8-7
8.4	Biokleptic use of Mms proteins for magnetite synthesis *in vitro*	8-10
8.5	Understanding Mms proteins *in vitro*	8-15
8.6	Development and design of additives: emergence of bioinspired magnetite nanoparticle synthesis	8-17
	8.6.1 Development from biomineralisation proteins: MmsF	8-17
	8.6.2 Screening non-biomineralisation proteins: magnetite interacting proteins	8-18
	8.6.3 Biomimetic magnetosomes	8-19
8.7	Summary: key learning, and the future (towards manufacture)	8-19
	References	8-21

9	**Case study 2: silica**	**9-1**
9.1	Biosilica occurrence and formation	9-1
9.2	Biomolecules controlling biosilica formation	9-4
9.3	Learning from biological silica synthesis: *in vitro* investigation of bioextracts	9-7
9.4	Emergence of bioinspired synthesis using synthetic 'additives'	9-8
	9.4.1 Which amino acids are important?	9-9
	9.4.2 Would (homo)polypeptides be sufficient to promote silica formation?	9-9
	9.4.3 Peptides from biopanning	9-11
	9.4.4 Do we need peptides or biomolecules?	9-13
	9.4.5 Can smaller molecules provide similar activities?	9-15
9.5	Benefits of bioinspired synthesis	9-16
9.6	From lab to market	9-18
9.7	Summary: key learning, summary and the future	9-22
	References	9-23

Preface

This book aims to provide an understanding of emerging bioinspired green methods for preparing inorganic nanomaterials.

Inorganic nanomaterials are used in many applications, ranging from sun cream to catalysis, as well as the latest innovations in nanomedicine and high density data storage. In recent years, we have understandably seen a large quantity of publication activity (including books) on the safety and toxicity of nanomaterials. However, there is a distinct lack of consolidated effort in addressing the sustainability of making nanomaterials. Current methods for nanomaterial synthesis are complex, energy demanding, multistep, and/or environmentally damaging, and hence clearly not sustainable. Green chemistry has great promise for future developments, particularly in sustainable designs for materials, processes, consumer goods, etc. However, to date, green chemistry has mostly focussed on the synthesis of fine chemicals and very rarely on nanomaterials.

New bioinspired/biomimetic approaches are emerging, which harness biological principles from biomineralisation to design green nanomaterials for the future. With reference to the significant body of research on understanding biomineralisation, Ozin *et al* state in their book, *Nanochemistry: A Chemical Approach to Nanomaterials*, that 'In molecular terms, it is relatively easy to comprehend the early stages of self-organisation, molecular recognition, and nucleation that precede the morphogenesis of biomineral form. It is not obvious however, how complex shapes emerge and how, in turn, they can be copied synthetically' [1]. In this book, the aim is to address this highly sought-after aspect of how to translate the understanding of biomineral synthesis into new green manufacturing methods. We cover areas from the discovery of new green synthesis methods all the way to considering their commercial manufacturing routes.

Who is the book for? The Royal Society of Chemistry and the American Chemical Society's Green Chemistry Institute have both highlighted a 'lack of a deep bench of scientists and engineers with experience in developing green nanotechnology' [2] as a significant barrier to the development and commercialisation of green nanotechnology. This has motivated us to write this book. When any of us have been educated within a specific traditional discipline of science or engineering for our undergraduate degree, it can be very daunting to take a leap into multidisciplinary science and study within the realms of new disciplines outside our comfort zone, where the experimental approach, culture and even language can be so different, creating barriers and challenges. However, the more we work at this interface, the more we realise that these boundaries are artificial, for the purpose of our education, and do not exist in Nature. The purpose of this book is to start with basic explanations to build a foundation, so that this area of science can become accessible to students from any related discipline. We hope that this book encourages scientists and engineers to become confident in bridging the gaps between chemistry, nanotechnology, biology, engineering and manufacturing. Specifically, the book combines green chemistry and nanomaterials in a single dedicated monograph.

As such, the book is written with a wide readership in mind, including primarily academic researchers focusing on synthetic biology and nanomaterials. It is targeted towards postgraduate students (taught and research degrees) undertaking studies pertaining to advanced materials and green, sustainable and/or environmental engineering or chemistry. Final year undergraduate students specialising in nanomaterials or green processes will also find this book valuable. Indeed, various universities currently run final year electives on nanomaterials, biomaterials, green chemistry, sustainability, etc, where this book is highly suitable as a textbook. Through the authors' interactions with industry, we know that many industries wish to learn more about these green technologies. Hence, we hope to reach industrialists and raise awareness of the emerging green manufacturing routes.

What is in the book? The book starts by introducing the principles of green chemistry and engineering (chapter 1). It then highlights the special properties that nanomaterials possess, their applications and ways of characterising them (chapters 2 and 3). It describes conventional methods of synthesising and manufacturing inorganic nanomaterials (chapter 4) and highlights that these techniques cannot always deliver the specifications required for applications or be sustainable (chapter 5). This will lead to the introduction of biological and biomimetic/bioinspired synthetic methods as a solution to precisely controlled nanomaterials as well as the design of sustainable manufacturing routes (chapters 6 and 7). The book elaborates on various mechanisms and examples of green nanomaterials (e.g. the role of an organic matrix and natural self-assembly, and advantages and opportunities with green nanomaterials). It will cover two case studies of magnetic and silica materials for advanced readers (chapters 8 and 9).

How to use the book. We acknowledge this book covers many different traditional disciplines and as such we cannot go into too much depth in every area. Furthermore, this is a very current and fast-moving research area. As new methods, materials and characterisation techniques are discovered, invented and developed, fairly recent advances become old quickly. For both reasons we recommend this textbook be supplemented with more detailed, specific and contemporary science and engineering research journal papers. Indeed, in the courses we teach on this subject, the material content of this book is used to explain the background and introduce current research papers as relevant examples.

A note on ongoing discussion on the topics covered in this book: In order to allow a dialogue between the readers, the authors and the publisher, we have created a dedicated web portal in order to receive feedback from readers and to allow authors and readers to post recent updates relevant to this book. This can be accessed at https://greennanobook.com/.

References

[1] Ozin G A, Arsenault A C and Cademartiri L 2009 *Nanochemistry: A Chemical Approach to Nanomaterials* 2nd edn (Cambridge: Royal Society of Chemistry), p 23

[2] Matus *et al* 2011 *Green Nanotechnology: Challenges and Opportunities* (Washington, DC: ACS Green Chemistry Institute)

Acknowledgements

We would like to acknowledge those who helped us complete this book. SP is sincerely grateful to Professor Steve Clarson and Professor Carole Perry who have been the sources of inspiration. SS would like to acknowledge Professor Andrew Harrison, Professor Neil Robertson and Professor Steve Evans for inspiring and enabling her to begin her research in this area. We thank the scientific communities to which we belong: the networks of academics we work and collaborate with and meet at conferences, where we share and develop new ideas, converse and debate. You are a constant source of inspiration for us and for science and engineering in this field to continue to develop. Our collaborators are acknowledged for sharing their wisdom and for the many stimulating discussions over the years. In particular, SS is grateful to Dr Bruce Ward and Professor Steph Baldwin for biological training and insight. We thank many of our current and past group members who have been instrumental in providing the ammunition for this book and for their patience during the writing stages. SS thanks Dr Andrea Rawlings and SP thanks Dr Joe Manning and Dr Mauro Chiacchia for their help with conceptualising some of the complex aspects/mechanisms included in this book. We are grateful to have had support from Ms Yung Hei Tung (Jodie), Dr Andrea Rawlings, Dr Johanna Galloway and Dr Scott Bird for artwork for some of the figures, and Ms Amber Keegan for help with copyright permissions. We also thank various funding agencies for supporting projects where we developed some of the ideas underpinning this book. In particular we thank EPSRC and BBSRC for funding the SynBIM project (EP/P006892/1, http://www.synbim.co.uk and BB/H005412/1). Finally, we thank the reviewers for their insightful feedback: from the initial book proposal, to friends providing comments on early drafts (thanks to Professor Maggie Cusack, Professor Marc Knecht and Dr Fabio Nudelman) and the reviewers of the completed draft. We offer sincere thanks to the publisher for their support and patience.

Finally, we would both like to thank our families. Academia is a challenging and intense career and this is only amplified when one chooses to write a book on top of our other commitments. We are most grateful to our families for their love and support both generally and specifically over the period of writing this book. We both have young children and are especially grateful: SP to his wife Geetanjali and SS to her husband Luke and our parents, for unquestioning childcare that enabled us to achieve this body of work. We are also grateful to our children: Ninaad and Nishaad, and Owen, Alex and Joel, for their interest in our work, for making us laugh and their inquisitive nature that reminds us every day what this is all for.

Siddharth V Patwardhan and Sarah S Staniland
Sheffield, August 2019.

Author biographies

Siddharth V Patwardhan

Siddharth is currently a Professor of Sustainable Chemical and Materials Engineering at the University of Sheffield. He obtained a first degree in chemical engineering at the University of Pune (India) followed by a master's and doctoral degrees in materials science at the University of Cincinnati (USA). He gained post-doctoral experience in inorganic chemistry at the University of Delaware (USA) and Nottingham Trent University (UK). After taking up a short-term lectureship in Chemistry, he became a Lecturer in Chemical Engineering at the University of Strathclyde in 2010. He then moved to Sheffield to take up a position of Senior Lecturer, where he was promoted to a Professor in 2018.

Siddharth leads the Green Nanomaterials Research Group (www.svplab.com), with a vision to develop sustainable routes to functional nanomaterials. His group focusses on the discovery of bioinspired nanomaterials, assessing their scalability and developing manufacturing technologies for energy, environmental, biomedical and engineering applications.

Siddharth is an EPSRC Fellow in Manufacturing and a Fellow of the Royal Society of Chemistry. He has played a key role in a number of national and international networks as well as conference organisation. One such symposium relevant to this book is on 'Green Synthesis and Manufacturing of Nanomaterials', as part of the ACS Green Chemistry and Engineering Conference in 2017. Siddharth is passionate about mentoring early career researchers and has received numerous awards including Dedicated Outstanding Mentor, Teaching Excellence and recognition as a *SuperVisionary* for all-round supervision.

Sarah S Staniland

Sarah is currently a Reader of Bionanomaterials in the Department of Chemistry at the University of Sheffield. She obtained an integrated undergraduate master's degree in Chemistry followed by a doctorate in Materials Chemistry (2001, 2005) both at the University of Edinburgh (UK).

After her PhD she won a prestigious independent EPSRC Life Science Interface Fellowship (2005–8) at the University of Edinburgh, where she initiated the research in which she is currently active. This helped her transition from chemical material sciences to interdisciplinary work at the interface with biology. She took this opportunity to live and work in various places globally, from Cape Town to Tokyo, forming lasting collaborations. She then took up a Lectureship in Bionanoscience in the School of Physics and Astronomy, University of Leeds in 2008, where she was promoted to Associate Professor in 2013. She moved to Sheffield in 2013 and was promoted to Reader of Bionanoscience in 2016.

Sarah leads the Bionanomagnetic Research Group which studies the biomimetic synthesis of magnetic nanomaterials, particularly inspired from how magnetite nanoparticles are produced within magnetic bacteria. From a basis of material chemistry and PhD in magnetic materials, Sarah has moved into a multidisciplinary approach of using biology to control material synthesis. She has been invited to speak at and organised national and international conferences to promote this research area and been a board member of the Royal Society of Chemistry Materials Chemistry Division. This multidisciplinary research field requires a highly skilled, open-minded and diverse research team, which she is passionate about training, developing and mentoring and is very grateful to them all. Sarah is committed to teaching, in particular multidisciplinary science which falls at the interface between several standard degree subjects, and is always experimenting with novel methods and techniques to improve her teaching in this area. She has taught a course on bionanoscience (covering much of the material in this book) for ten years. Sarah has won three prestigious awards recently, including two for her research: the acclaimed RSC Harrison Meldola Award in 2016 and the Wain Award in 2017, as well as the Suffrage Science Award in 2017 for her work on the promotion of gender equality.

Section I

Green chemistry principles

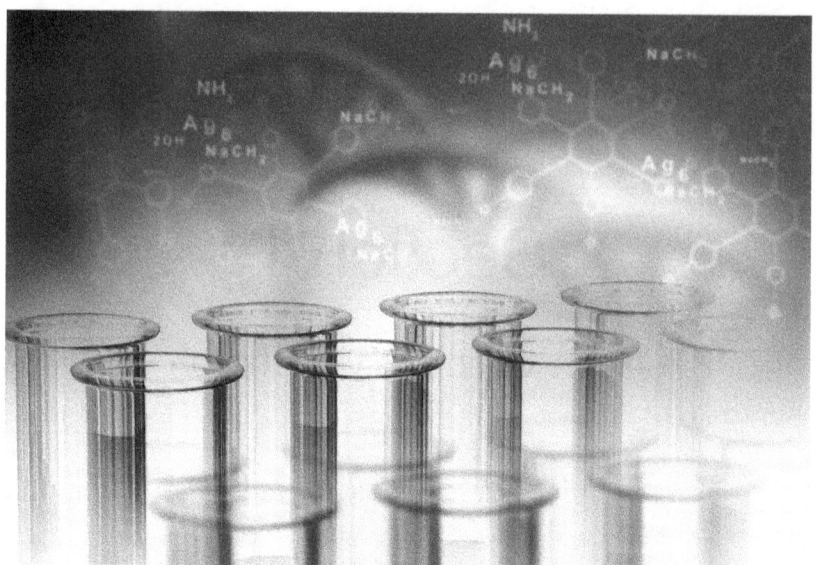

Image courtesy of bluebay/Shutterstock.

This short section consists of a chapter on green chemistry and engineering. It introduces the 12 principles of green chemistry and various drivers for making a given process or product greener. Further, ways to improve sustainability are discussed mainly in terms of the cost of the waste produced. A brief introduction is provided on how to evaluate the sustainability or green credentials of a given process or product leading to a discussion on ways to improve sustainability. These concepts will be used in other chapters in the book in order to explore potential (un)sustainable aspects of a given method for nanomaterials synthesis. This section aims to set the scene for the book and the principles explained will be revisited in later sections of the book, in order to put them in the context of nanomaterials synthesis and manufacturing.

IOP Publishing

Green Nanomaterials
From bioinspired synthesis to sustainable manufacturing of inorganic nanomaterials
Siddharth V Patwardhan and Sarah S Staniland

Chapter 1

Green chemistry and engineering

1.1 Principles of green chemistry and engineering

1.1.1 Overview

At present, we are highly reliant on chemical processes because of the benefits they bring to us. Examples of sectors where chemical industry has impacted our lives include healthcare, transportation, communications and food, constituting total sales of around €1.5 trillion globally [1]. In recent times, society has become more aware and concerned about the negative environmental impact associated with chemical industry. As such, there have been wider legislative interventions in order to ensure greater control and monitoring of chemical processes. These changes have led to increasing pressure on the chemical industry to continue to deliver high value products in an economical fashion, while minimising or eliminating the adverse environmental burden. This new challenge has driven the development of green chemistry for sustainable chemical processing.

Sustainable development has been defined by the United Nations as '…meeting the needs of the present without compromising the ability of future generations to meet their own needs' [2]. Further, the United States Environmental Protection Agency (EPA) has extended this definition to give birth to *green chemistry* where the main goal is to 'promote innovative chemical technologies that reduce or eliminate the use or generation of hazardous substances in the design, manufacture and use of chemical products' [3]. Related to this is *green engineering*, which pertains to the design, commercialisation and the use of processes and products in an economical fashion to minimise pollution and risks to health and the environment. Utilising these concepts, 12 principles of green chemistry have been formulated [4], which are shown in figure 1.1.

Applying these green and sustainable principles has the potential to address the challenges facing chemical industry, in balancing the high value of products against the environmental burden. A common perception is that green chemistry is costly and/or can lead to costly products, thus reducing the profit for industry. However,

> **1. Prevention**
> It is better to prevent waste than to treat or clean up waste after it has been created.
>
> **2. Atom economy**
> Synthetic methods should be designed to maximize the incorporation of all materials used in the process into the final product.
>
> **3. Less hazardous chemical syntheses**
> Wherever practicable, synthetic methods should be designed to use and generate substances that possess little or no toxicity to human health and the environment.
>
> **4. Designing safer chemicals**
> Chemical products should be designed to affect their desired function while minimizing their toxicity.
>
> **5. Safer solvents and auxiliaries**
> The use of auxiliary substances (e.g., solvents, separation agents, etc.) should be made unnecessary wherever possible and innocuous when used.
>
> **6. Design for energy efficiency**
> Energy requirements of chemical processes should be recognized for their environmental and economic impacts and should be minimized. If possible, synthetic methods should be conducted at ambient temperature and pressure.
>
> **7. Use of renewable feedstocks**
> A raw material or feedstock should be renewable rather than depleting whenever technically and economically practicable.
>
> **8. Reduce derivatives**
> Unnecessary derivatization (use of blocking groups, protection/deprotection, temporary modification of physical/chemical processes) should be minimized or avoided if possible, because such steps require additional reagents and can generate waste.
>
> **9. Catalysis**
> Catalytic reagents (as selective as possible) are superior to stoichiometric reagents.
>
> **10. Design for degradation**
> Chemical products should be designed so that at the end of their function they break down into innocuous degradation products and do not persist in the environment.
>
> **11. Real-time analysis for pollution prevention**
> Analytical methodologies need to be further developed to allow for real-time, in-process monitoring and control prior to the formation of hazardous substances.
>
> **12. Inherently safer chemistry for accident prevention**
> Substances and the form of a substance used in a chemical process should be chosen to minimize the potential for chemical accidents, including releases, explosions, and fires.

Figure 1.1. 12 principles of green chemistry, after [4]. Copyright of OUP 1998.

this is not always true. While there may be initial investment needed for the development of green technologies, which can replace existing processes, such changes can also lead to reduced costs of production and products. Besides, considering the long-term benefits of adopting green chemistry, initial costs can be offset. Some of these points and the 12 principles will be discussed further in chapters 4 and 5 with specific relevance to nanomaterials.

1.1.2 Drivers for green approaches

One of the main issues leading to adverse environmental impact is the generation of waste in a chemical process. Depending on the technological, economical and legislative frameworks, the quantities of waste produced, in particular, in relation to the amount of product produced, vary from sector to sector. For example, some industrial sectors are technologically very advanced, such that they have developed ways to minimise waste, e.g. petroleum refining. On the other hand, in some other sectors, the cost of the product is significantly higher than the loss of value from waste or cost of treating waste, and hence waste minimisation is not given due importance (e.g. pharmaceuticals). There are also examples where pro-active legislation has driven the chemical industry to find innovative ways to minimise waste. Waste relates to inefficiencies in a given process, which leads to loss of valuable resources (e.g. substances and energy) and it can cause risks to the environment and health, which ultimately increases the process costs. Figure 1.2 illustrates the financial, environmental and societal origins of the costs associated with waste.

If one considers the total costs for waste originating from all areas as shown in figure 1.2, it implies that in order to drive green chemistry, the process or product costs associated with a new green technology (N) should be lower than or at least equal to the costs of the existing (wasteful) processes (E),

$$\text{i.e. } N \leqslant E. \tag{1.1}$$

Along similar lines, the profits from the new process should be higher or at least the same when compared to the existing process. The cost of green technology includes new production costs (P') and the investment needed for developing new green technologies (G), which would reduce the waste and avoid the loss of the image of the business. On the other hand, continuing to operate using existing process

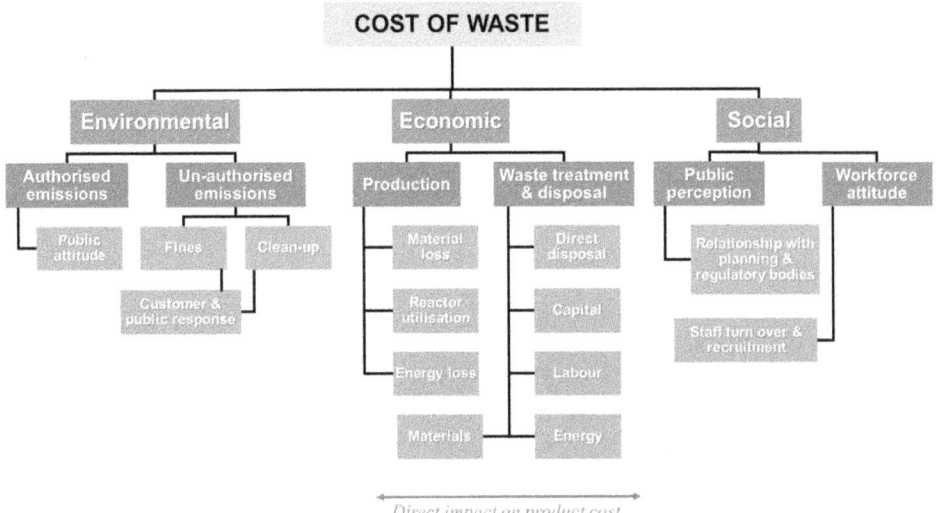

Figure 1.2. Costs associated with waste. Reproduced with permission of the Royal Society of Chemistry from [1].

includes current production costs (P) and incurs waste management costs (W) as well as costs associated with the loss of public image (I),

$$\therefore G + P' \leqslant P + W + I \tag{1.2}$$

$$\therefore G \leqslant (P - P') + (W + I). \tag{1.3}$$

This is a simplistic approach to describe various scenarios and consequences. First, if the company does not wish to invest at all ($G = 0$), then it means that the new process costs can be higher than the existing process costs, but only by the sum of the costs of waste management and those associated with the loss of image,

$$\text{i.e.} \therefore (P' - P) = (W + I). \tag{1.4}$$

Without any investment, it is not easy to obtain access to a new and sustainable process. Further, if the costs for managing the changes in the process (e.g. downtime, marketing, customer satisfaction, etc.) are considered, then in reality,

$$\therefore (P' - P) \ll (W + I). \tag{1.5}$$

This leaves very little incentive for developing a greener process and highlights the strong need for initial investment and the motivation for sustainability at all levels of the organisation. On the other hand, if a new process is developed such that there are no additional process costs ($P - P' = 0$), then the investment needed would be of the order of the cost savings from waste reduction and maintenance of the image,

$$\text{i.e.} \ G \leqslant W + I. \tag{1.6}$$

The production costs and the type and amount of waste generated are not just dependent on the chemical reaction. Other factors such as choice of solvent and downstream separation and purification processes play a major role. Therefore, developing greener approaches is not about simply operating at lower temperatures or using volatile solvents, for example. Sometimes it may be about balancing the priorities or perhaps radically changing the processes.

1.1.3 Estimating environmental impact

In order to make informed decisions about the need for green innovations for a given process, it is important to assess the environmental impact of that particular process. There are various methods and tools available to qualitatively and semi-quantitatively analyse the environmental impact of processes [1]. Selected methods are described below with their principles, use, advantages and limitations.

Environmental factor (E-factor) [5, 6], also known as waste-to-product ratio (equation (1.7)), is a simple measure for identifying the amounts of waste/by-products produced with respect to the mass of the product.

$$\text{E-factor} = \text{mass of waste and by-product} \div \text{mass of products.} \tag{1.7}$$

Sheldon and co-workers [5] have studied E-factors for various industries and reported great variations from one industrial sector to another (table 1.1).

Table 1.1. E-factors for various sectors. Adapted from [5].

Industry sector	Product capacity, tonnes	E-factor
Oil refining	10^6–10^8	~0.1
Bulk chemicals	10^4–10^6	<1–5
Fine chemicals	10^2–10^4	5–50
Pharmaceuticals	10–10^3	25 to >100

The main reasons for such variations appear to be related to the product value/profit margins, relevant legislations, cost of waste and market competition. The E-factor analysis of a process is simple and provides quick estimates of wastefulness, which can lead to waste minimisation campaigns. While E-factor is easy to use, it can provide misleading information in some cases. For example, consider the following reaction:

$$\textbf{Reactant 1 + Reactant 2} \xrightarrow{Solvent\ A} \textbf{product + waste}. \quad (1.8)$$

In reaction (1.8), when calculating the E-factor, if water is the waste, it will be treated as any other waste, although water is not inherently toxic or hazardous. In order words, E-factor does not take into account the nature and the actual impact of the waste or by-products and can treat waste on equal grounds despite significantly different environmental impacts.

Environmental quotient (EQ) [5] has been introduced in order to address the weakness of E-factor. EQ is essentially a modified E-factor, which takes into account the environmental 'unfriendliness' quotient (Q) of the waste or by-products (equation (1.9)). **Effective mass yield** (EMY) is another similar metric (equation (1.10)), which also disregards benign substances used or produced. Therefore, both EQ and EMY help to distinguish between hazardous waste and non-hazardous waste. However, both metrics are susceptible to inconsistencies due to the vagueness around what is environmentally unfriendly or benign, leading to debate over what values to assign to individual substances.

$$EQ = \text{E-factor} \times Q \quad (1.9)$$

$$\%EMY = 100 \times (\text{mass of products} \div \text{mass of non-benign materials used}). \quad (1.10)$$

Life cycle analysis (LCA), is an extensive way of assessing the environmental impact and sustainability of a given process or product. LCA analyses the entire life cycle of a product, from the extraction of raw materials all the way to the fate of the product. In the context of nanomaterials, this has been extensively reviewed elsewhere [7]. This comprehensive analysis can overcome various pitfalls associated with the other metrics introduced above. However, performing LCA is laborious and time consuming, and it requires a large amount of process- and product-related data to be available. One important advantage of LCA is that for an alternative process or product, LCA can help differentiate between pollution/waste prevention and shifting pollution. For example, consider the reaction (1.8) shown above, and assume that the solvent A is a hazardous solvent. In order to remove the need for solvent A, an alternative reaction is available using a different precursor (equation (1.11)), where solvent B is benign:

$$\text{Reactant 3} \xrightarrow{\text{Solvent B}} \text{product} + \text{waste}. \quad (1.11)$$

On the face of this new reaction, it appears to be 'green' because the hazardous solvent has been replaced with a non-hazardous one. However, during LCA, one needs to consider how the new reactant is derived (reaction (1.11)). Solvent C (equation (1.12)) used to produce reactant 3 may be as hazardous as solvent A. It quickly becomes clear that the alternative (reaction (1.11)) is not as green as it appeared.

$$\text{Reactant 1} \xrightarrow{\text{Solvent C}} \text{Reactant 3} + \text{waste}. \quad (1.12)$$

Essentially, although the need for solvent A has been removed, in order to produce the same product, solvent C is required, thus simply shifting the problem to the synthesis of the new precursor. In addition, LCA can take into account the energy required to produce the desired products.

Carbon footprint is another measure for assessing the environmental impact of a process or a product. It estimates the CO_2 equivalent emissions of greenhouse gases caused by a process or associated with a product. Carbon footprint analysis can help decide between processes where, for example, one option can produce a high quality product but require very high amounts of energy, while an alternative method could be lean on energy requirements but may produce a lower quality product.

1.2 Ways to improve sustainability

Estimation of environmental impact using the metrics discussed above can help identify the problems with an existing process or product. This can then help device targeted plans to find alternatives. Environmental impact assessment can also highlight future challenges, such as the need for technical developments in order to improve product quality for greener processes such that they can be competitive.

There are many ways to improve process sustainability, and most of them are based around finding alternatives to either the solvents used, the reactants used and/or the process conditions (e.g. temperature or pressure) [1]. Avoiding the use of solvents altogether is an excellent example of improving greenness of a process. Other alternatives include switching to solvents that are non-volatile organic compounds. The use of supercritical fluids (e.g. CO_2 and water) has also been reported as an alternative due to the ease of solvent separation and reuse. However, one should consider the energy required to operate processes under supercritical conditions (typically high temperatures and pressures; e.g. for water, the critical point is 374 °C and 220 bar). In the case of processes that require heating, alternative ways of providing energy, such as microwave or ultrasound, could be effective. These methods work on the principle of providing energy only to the desired location or chemicals without wasting energy by heating the entire medium or system.

Various ways of estimating the environmental impact, as discussed above, are powerful tools, however they require involvement and inputs from chemists and process engineers [7]. It is critically important to analyse how the modifications made to a section of a process can affect the entire process and perhaps beyond to the entire business. For example, environmental impact estimates cannot predict how changes in feedstock can affect the product value, profits and market competition due to feedstock availability. Answers to such questions can be obtained by performing techno-economic evaluation of a process [8].

1.3 Green chemistry and nanomaterials

When compared to the sectors listed in table 1.1, nanomaterials manufacturing is relatively young. Most attention has been focussed on the discovery and design of nanomaterials. Although the use of nanomaterials can be back-dated to the 9th century (see chapter 2, section 2.2), the large-scale commercial production of engineered nanomaterial developed and accelerated around the late 1990s and early 2000s [9]. This means that the manufacturing-related developments for nanomaterials are in their infancy, and for a vast majority of newly discovered nanomaterials, large-scale manufacturing does not exist. In other words, we are *happy* to have *some* process for the large-scale manufacturing of desired nanomaterials, while little or no attention is given to the sustainability of the process. However, this needs to change and the time is ripe to focus on the environmental impact of nanomaterials production, in order to apply the developments enjoyed by other sectors, listed in table 1.1, to nanomaterials production. These factors clearly stress the urgent need for developing fundamentally new production methods for nanomaterials that are green and sustainable. This change in thinking is very important, and previous experience suggests that when sustainability/green principles are included at the discovery stage, this provides the most benefits.

This book focuses on stimulating a shift towards a sustainable approach for nanomaterials—from discovery to production. Chapter 2 will serve as an introduction to nanomaterials. Their properties will be discussed and benefits offered from their use in selected applications will be presented. In chapter 3, we will provide an overview of the analytical techniques used to probe various properties of nanomaterials. Chapter 4 will present, with examples, a range of current manufacturing processes used to produce nanomaterials at large scales. We will discuss these processes for sustainability by using the theory and concept of green chemistry covered in chapter 1. Chapter 5 will summarise the benefits of using nanomaterials, while highlighting the need for greener alternatives. In chapter 6, we will detail how biology produces a range of inorganic materials from macro to nanomaterials and point out the potential for learning from biology. This learning will be consolidated in chapter 7 where strategies for developing biologically inspired green routes to produce nanomaterials will be presented. The key advantages and opportunities from these alternative routes will be identified, and further explained in chapters 8 and 9 with two case studies.

References

[1] Lancaster M 2010 *Green Chemistry: An Introductory Text* 2nd edn (Cambridge: Royal Society of Chemistry)
[2] Gro Harlem Brundtland 1987 Development and international co-operation: environment, the United Nations, A/42/427 1987, http://www.un-documents.net/wced-ocf.htm
[3] Basics of Green Chemistry [cited 2016 23rd March]; Available from: https://www.epa.gov/greenchemistry/basics-green-chemistry
[4] Anastas P T and Warner J C 1998 *Green Chemistry: Theory and Practice* (Oxford: Oxford University Press)
[5] Sheldon R A 1997 *J. Chem. Technol. Biotechnol.* **68** 381
[6] Sheldon R A 2007 *Green Chem.* **9** 1273
[7] Windsor R, Cinelli M and Coles S R 2018 *Curr. Opin. Green Sustain. Chem.* **12** 69
[8] Drummond C, McCann R and Patwardhan S V 2014 *Chem. Eng. J.* **244** 483
Lieberman M B 1989 *Strat. Manage. J.* **10** 431
[9] Vance M E, Kuiken T, Vejerano E P, McGinnis S P, Hochella M F, Rejeski D and Hull M S 2015 *Beilstein J. Nanotechnol.* **6** 1769

Section II

Nanomaterials

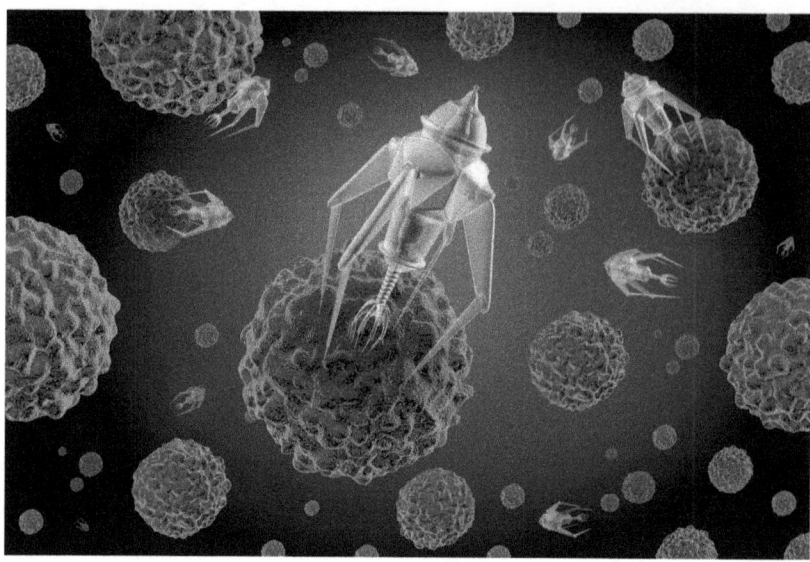

Image courtesy of Lightspring/Shutterstock.

We have learnt the benefits of green synthesis, but before we can explore sustainable bioinspired green synthesis of nanomaterials, it is important to understand what nanomaterials are, and how we can use them. Nanomaterials, nanotechnology and particularly biomedical applications of nanotechnology have captured our imagination in both science and fiction for decades, even before nanotechnology was scientifically realised: the concept of nanobots, removing diseased cells from our blood stream, is a very clear image in sci-fi literature, and excitingly it is actually something that we believe will become a reality in the coming years.

This section is devoted to highlighting special properties of nanomaterials within the scope of this book, illustrating their functions and how these properties serve several emerging and well-known applications with selected examples. Given these special features, which may not be found in bulk materials, an entire chapter focuses on giving an introduction to various key characterisation techniques used for studying and measuring properties of nanomaterials. This is followed by a description of 'top-down' and 'bottom-up' current and emerging methods for nanomaterials synthesis and manufacturing along with the physicochemical principles of nanomaterial formation. Suitable examples of the chemistry of materials are provided to help illustrate the processes.

IOP Publishing

Green Nanomaterials
From bioinspired synthesis to sustainable manufacturing of inorganic nanomaterials
Siddharth V Patwardhan and Sarah S Staniland

Chapter 2

Nanomaterials: what are they and why do we want them?

2.1 Fundamentals of the nanoscale

Although the term 'nanotechnology' is commonly used beyond science by the general public and in the media, an understanding of what 'nano' is defined in length scale, and the changes to physical properties that occur in materials in this miniaturised world, are not generally realised.

'Nano' comes from the Greek word for 'dwarf' and is the prefix of a measurement that is $\times 10^{-9}$ (or one billionth of that unit). In the context of material science and nanomaterials we are interested in the measurement unit of length: metres, and therefore the nanometre (nm), although 'nano' can prefix any unit, for example nanosecond, nanogram or nanomole. Put into context, there are one million nanometres in a millimetre and a billion in a metre. Figure 2.1 may be used to aid visualisation of these different length scales. Considering it from the opposite side, scaling up rather than down, we measure atomic bond lengths in Angstroms (Å) which are $\times 10^{-10}$ m, and so the nanometre length scale is 10× larger than the length scale at which we consider atomic reaction (chemistry) to occur. For example, the unit cell (smallest crystal unit) of sodium chloride is 0.56 nm^2, containing four of each type of (sodium and chloride) atoms. Although performing chemical reactions at this scale is an old and well-established discipline, the idea of crafting materials and components on the nanoscale and designing and creating tailored materials for smart applications is now developing as the relatively new disciplines of nanoscience and nanotechnology.

In 1959, Professor Richard Feynman delivered his seminal lecture entitled 'There's plenty of room at the bottom', in which he imagined what could be achieved if we shrunk technology down to the nanoscale. This lecture inspired material scientists and technologists to challenge themselves to venture to the smallest length scales possible for materials. The lecture has thus been defined as the birth of nanoscience and nanotechnology. As a physicist, Professor Feynman was fully aware that this shrinking

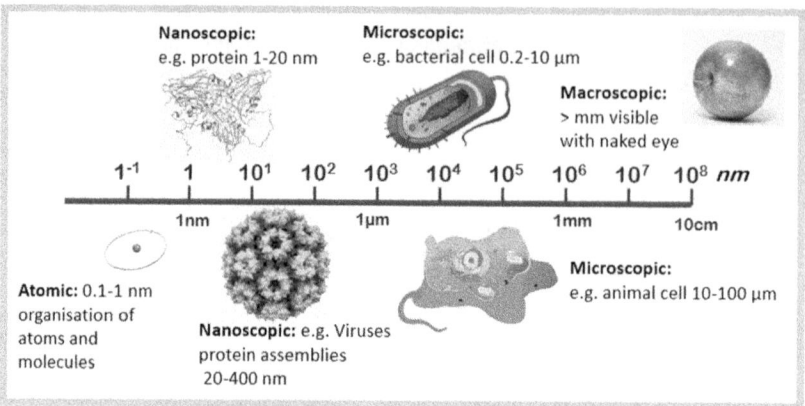

Figure 2.1. Schematic of length scale with respect to a nanometre, showing common examples. Source: Abhijit Tembhekar/LadyofHats/Alan Cann/Acharya KR, Fry E E, Logan D T, Stuart D I/; visualisation: Astrojan/ Белых Владислав Дмитриевич.

would not be a simple linear scaling issue, because smaller components will not display the same properties as the bulk material. Rather, materials at the nanoscale display rich and complex properties specifically dependent on their dimensions, proportion of surface atoms, and morphology (see the examples in the sections below).

Nanomaterial can be defined as having at least one dimension in the nanoscale (between 1–100 nm). This ranges from thin nanometre films (one dimension at the nanoscale) to nanowires (two dimensions at the nanoscale) to nanoparticles (all three dimensions at the nanoscale). Dimensions smaller than 1 nm are in the territory of clusters, made up of a few atoms, which are outside the realm of nanomaterials. We should also consider that the nanoscale can apply to voids as well as matter, and thus nanoporous materials can also be considered as nanomaterials, particularly as they have large surface areas. While 100 nm is conventionally the upper limit, this can be blurred. Materials termed nanoscale are reported over this size, if the properties displayed are still within the realms of nanoproperties (discussed in section 2.2) or, (more frequently as time goes on) it suits to classify them as a nanomaterial for political (non-scientific) reasons!

The nanoscale is a most interesting length scale for solid state materials for several reasons. First, from a philosophical viewpoint, it can be asked: how small can a particle be before it is no longer a material? Indeed, there is a valid scientific debate about whether or not a nanoparticle can or cannot dissolve, or if it is already a solute, when in a solution.

Second, nanomaterials have a large percentage of surface compared to bulk material, and surfaces have very different chemical and physical properties to the bulk. To illustrate this, let us consider two spherical particles with radii of 1 μm and 10 nm. The ratio of the external surface area to volume ($A:V$) of each particle is inversely proportional to its radius (figure 2.2):

$$\frac{A}{V} = \frac{4\pi r^2}{\frac{4}{3}\pi r^3} = \frac{3}{r}. \tag{2.1}$$

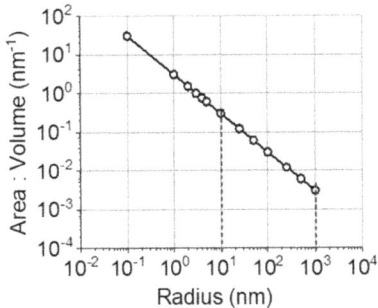

Figure 2.2. Graphical representation of how surface area increases with decreasing radius of a spherical particle (on a log 10 scale).

For these two particles, $A{:}V$ will be respectively 0.003 nm^{-1} and 0.3 nm^{-1}. This means that the 10 nm particle will have 100 times more surface for the same volume. As a result, the 10 nm particle will have 100 times more atoms on the surface available to interact with the external environment. This phenomenon leads to significantly increased catalytic activities, for example. This effect also manifests into other special effects and properties such as optical and magnetic properties, which are discussed further below.

Third, the nanoscale lies at the boundary of bulk material properties and atomic chemical properties (figure 2.3). The former is generally described by continuous transitions, while the latter must be considered as quantum mechanical. Quantum mechanics is both difficult and abstract to conceptualise but can be linked to the macroscale world by the theory of wave particle duality, where any wave can also be described as a particle with the same energy, and vice versa.

$$E = hf = pc. \quad (2.2)$$

These are linked by the work of both Einstein and Broglie, that shows that a particle and wave with the same energy are interchangeable (equation (2.2), where E = energy, h = Planck's constant, f = frequency of the wave, p = particle momentum = mass × velocity, (momentum of a particle) and c = the speed of light). At this stage however, it is simple to see that properties at the atomic scale are quantised because electrons (the principle particle (or wave) behind all chemistry) reside in discrete quantised electronic orbitals (defined by having an integer wavelength). This is simply described as if one considered the energy a particle can have in a defined narrow box, which is the energy that equates to any integer wavelength that will fit in the box as a standing wave (figure 2.3). These energy levels are equal distance apart as the box width remains constant (figure 2.3), however, the energy well of an atom follows a curve and as such the energy levels become closer together the higher the energy (figure 2.3). This is also directed by the shape of the energy well; the deeper the well, the further apart the energy levels, while the value of each energy level is also directed by the element (the pull of the nucleus etc). Thus, there is only one series of discrete energies that relates to that electron and its energy levels, of that particular atom. As the number of atoms in a material increases, so too does

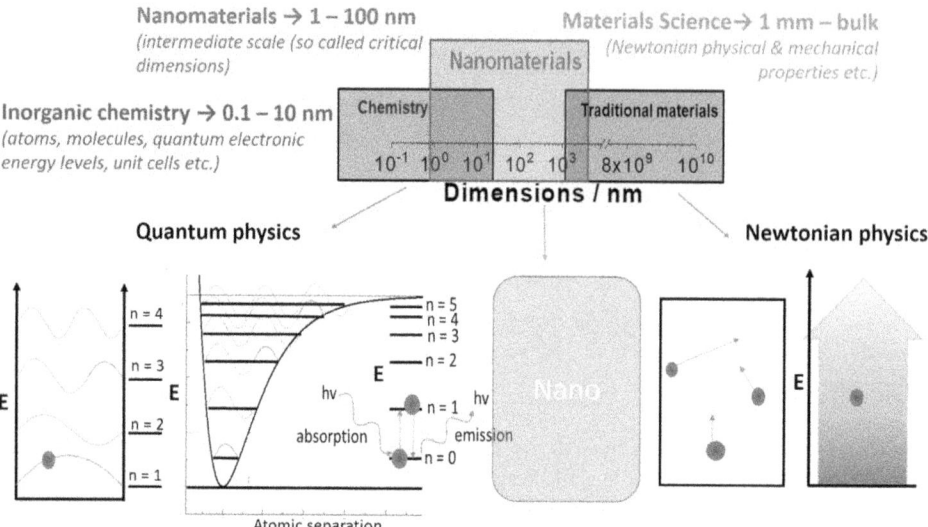

Figure 2.3. Depiction of how nanoscale materials lie at the crossover between the atomic quantum laws of physics and the macroscale Newtonian laws of physics. In quantum physics: the left shows how a particle in a very narrow box can only have the energy quantised by the size of the wavelength. The change in energy of these is the same as the box sides are parallel. For an atomic system, the change in energy between levels decreases at increasing number of n, due to the energy potential. Conversely, a particle in a macroscale box can move in any direction at any given energy, so there is a continuum of energy a particle can possess.

the number of interactions between atoms and electrons. The more interactions, the harder it is to define discrete energy levels as more and more energy levels occur, forming a continuum which is what we can describe in a bulk material (figure 2.3). One can similarly conceptualise the particle in a box on a macroscale. The width of the box is now much greater and the wavelength associated with a macroscale (large mass) particle is now so small, the particle can now be considered to occupy any energy, so it becomes a continuous system. In Newtonian physics a particle can have any energy, velocity and thus momentum within its box (figure 2.3). The interface where quantum states transition to continuous phases is at the nanoscale and this leads to a range of exotic behaviours, some of which are briefly described below in section 2.3.

2.2 Tangible and historical examples of nanomaterials

It is no coincidence that nanoscience has grown in parallel with the growth and development of instrumentation, imaging and spectroscopic techniques that have the resolution to allow probing on the nano- to the atomic scale (see characterisation in chapter 3). This has been essential to analyse and understand these materials as well as aid design and synthesis. However, while we may not have been able to visualise or understand them, we have been using the properties of nanomaterials for millennia, with examples stretching through history (figure 2.4). One of the earliest examples is the 4th century Roman Lycurgus cup. This beautiful glass cup, mounted

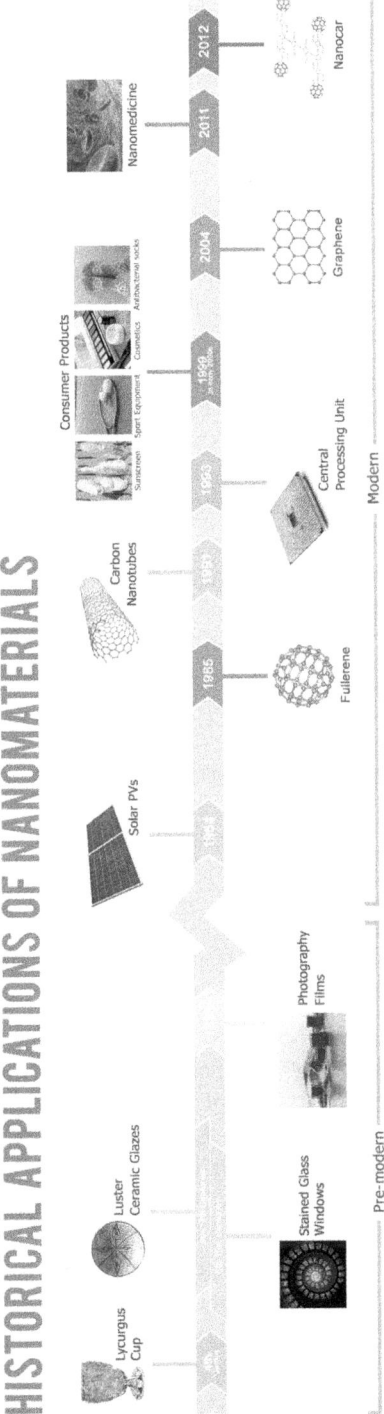

Figure 2.4. Schematic of nanoscale materials and accompanying timeline of their development.

in a carved bronze exterior, is now housed in the British Museum. It is an excellent example of nano-optics (specifically surface plasmon resonance), where, because it contains different metallic nanoparticles it looks green if illuminated from the front, because silver nanoparticles scatter the reflective light, and red if illuminated from behind, due to surface plasmon resonance from the gold nanoparticles interacting with transmitted light. It was not until 1847 that Faraday first reported nanoparticles; gold nanoparticles, as a colloidal suspension, he called a 'gold sol'. He was the first to recognise and describe the exotic optical properties gold nanoparticles possess, which is perhaps the first recognition of quantum size effects and thus quantum mechanics. Not quite so complex but equally dramatic and awe-inspiring, stained glass windows have been used for centuries in religious buildings. Again, metallic nanoparticles are responsible for the incredible colour still enjoyed today in these and also in ceramic glazes, centuries after they were created. The photo sensitivity of silver was developed during the 1800s to invent photography and film (the origin of the term 'silver screen') as we knew it up until the late 20th century. While digital imaging has now overtaken film to become the norm, many professionals still prefer traditional film for its distinctive look, and also for its better resolution. Colloid silver on film has nanoscale resolution! The later 20th century saw the explosion of nanomaterials, both in their own right (the subject of this book) and for inclusion in many consumer products, such as solar photovoltaic materials in solar panels and tyres. We even now see silver nanoparticles embedded into socks to for their antimicrobial properties, to stop them smelling! In the present day, our understanding of nanoproperties and ability to make and characterise nanomaterials is growing, and so too is our ambition for their use. The field of nanomedicine has been explored in recent years, and is the subject of section 2.4.1. While curing disease is one of the key applications, it is also the aim of nanoscientists globally to create the smallest working nanomachines, using classic engineering shapes such as axials, rings and wheels at a molecular scale, to make nanowalkers, molecular motors and nanocars [1].

2.3 Special properties offered by the nanoscale

There are many physical properties that are affected at the nanoscale, such as melting point and electronic properties. Furthermore, due to the large surface area and predominance of surface chemistry, nanomaterials are also ideal for heterogenous catalysis. The purpose of this chapter is NOT to give a complete, definitive account of all properties and applications of nanomaterials, but rather to give a flavour and thus must focus on a small selection of nanoproperties and their subsequent application.

There are two features of nanoscale materials: (1) increased surface-to-bulk volume ratio and (2) the reduction in size, which leads to the surface and quantum mechanical properties predominating over the bulk properties. Both key features are beautifully illustrated in turn by considering two optical properties of nanomaterials: the surface effect of surface plasmon resonance (SPR) and the quantum mechanical effect of quantum dot fluorescence. Furthermore, the reduction in size also affects magnetic properties. These three nanoproperties will be described in this section, with the following section showing real-world applications of these specific properties.

2.3.1 Optical: surface plasmon resonance

A surface plasmon is the interaction of the free surface electrons of a metal with electromagnetic radiation. Surface plasmons are mostly seen in inert noble metals, but all metals and even some highly doped semiconductors will exhibit a surface plasmon effect. When a metal surface is irradiated with light of a larger wavelength than the size of the particle, the conducting electrons will polarise and collectively oscillate about the surface of the particle in response to the wave, and are thus driven to form a resonance standing wave, giving a resonance peak (figure 2.5(i)). This is

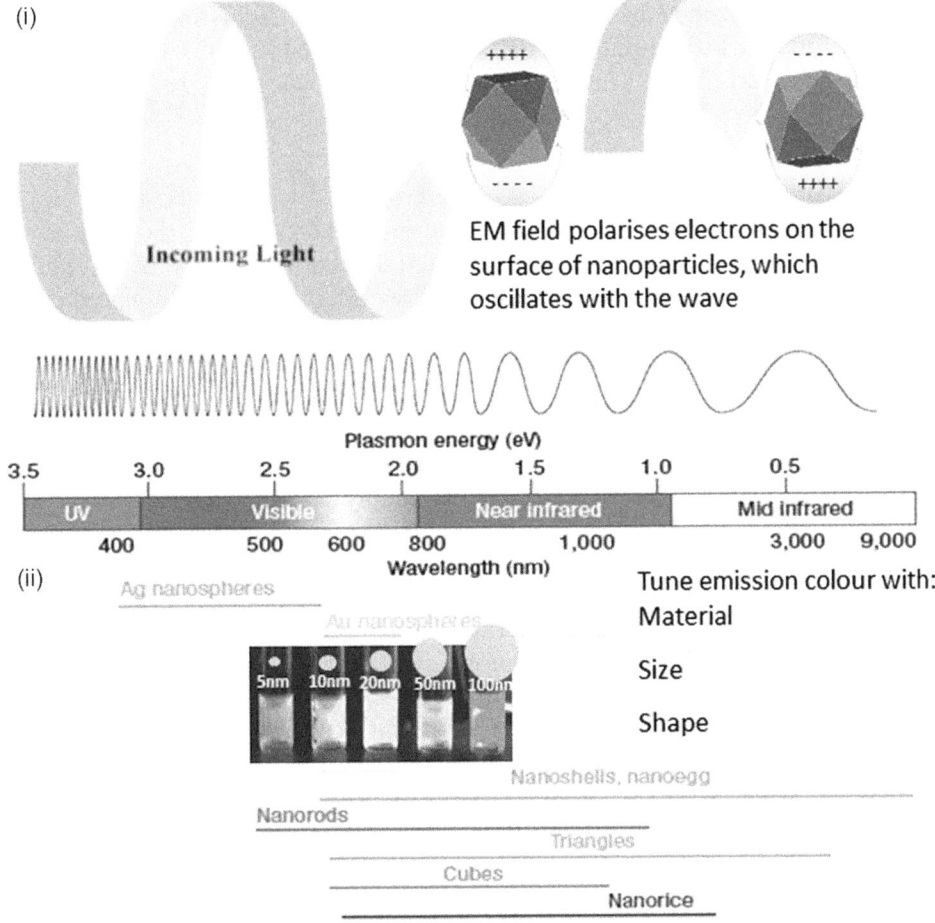

Figure 2.5. (i) Schematic of how electromagnetic wave interacts with surface electrons on a nanoparticle (adapted from figure in [2] with permission of the Royal Society of Chemistry). (ii) Regions of the electromagnetic spectrum that specific nanomaterials of different morphologies emit due to SPR (adapted from figures in [3] with permission of Springer).

known as SPR. At this peak the electromagnetic radiation is both scattered and absorbed by the metal [2]. Some of the radiation is absorbed by the nanomaterial and converted to crystal lattice vibrational energy (phonons), while the rest of the radiation is emitted in different directions (scattering).

The colour of the light emitted comes predominantly from the scattering, so we should consider what contributes to the SPR scattering. We need to think about polarisability; the ease with which charges (the conduction electrons) on the nanoparticle surface undergo charge (re)distribution and form partial dipoles. This is dependent on how far the electrons need to move, i.e. the cross-sectional area (so in turn, the radius of the particle (r)). It also depends on the nanoparticle material, and how easy electrons can move on this metal: the dielectric function of the nanoparticle ($\varepsilon = \varepsilon_1 + i\varepsilon_2$) (wavelength dependent) and the dielectric function of the medium (ε_m) in direct contact with the surface. This can be quantitatively described by the Gustav Mie theory of light scattering for any spherical particles with radius = R, for which a simplified version (when $2R \ll \lambda$) is shown in equation (2.3)

$$C_{\text{ext}} = \frac{24\pi^2 R^3 \varepsilon_m^{3/2}}{\lambda} \cdot \frac{\varepsilon_2}{(\varepsilon_1 + 2\varepsilon_m)^2 + \varepsilon_2^2}. \tag{2.3}$$

Here C_{ext} is the extinction cross-section and gives the quasi-polarizability of the nanoparticle. SPR is a property of nanomaterials due to the scale of wavelength verses particle size, and the increased surface area offers a more pronounced effect. Furthermore, the wavelength of the resonance peak and thus the colour emitted by the SPR is dependent on the components highlighted above, the primary ones being the particle composition (material), and the radius. However, what equation (2.3) does not describe is the effect of shape/architecture, as the Gustav Mie equation is only for a simple spherical particle. Surface plasmons are highly dependent on the overall surface architecture, and not limited simply to size. Varying morphology affects the SPR frequency and thus emission since varying the shape alters the surface architecture and this alters the electric field distribution on the surfaces, again altering the surface plasmon resonance. This tends to result in an increase in the range of resonance frequencies; this is certainly true for cubes, rods, cages/shells and egg shapes (figure 2.5(ii)). Furthermore, all other volumes have a larger surface area to volume ratio than a sphere, so all other shapes will offer increased surface area. This is particularly true for hollow shapes like cages.

The classic historical example is gold nanoparticles. Spherical gold shows a range of colour dependence on size, from green for particles around 5 nm, up to red for particles of about 50 nm (figure 2.5(ii)). However, silver spherical nanoparticles have higher frequency SPR resonance (higher energy), emitting from green up to ultra-violet compared to gold nanoparticles of the same size and shape (gold emitting red to green), demonstrating the dependence SPR has on material as well as the size. Finally, since this is a surface effect and the surface is interacting with the surrounding media, we cannot forget the SPR frequency dependence on the dielectric constant of the media. Furthermore, SPR is affected by the locality of near materials exhibiting SPR [4]. Both are incredibly useful for applications as colorimetric indicators/sensors as very subtle changes to either the nanoparticle surface, the surrounding media, or the distance nanoparticles are from each other can be optically detected. In addition, SPR is so intense that the scattering of a single nanoparticle can be seen by the eye in an optical scattering microscope. The blue light emitted by a silver nanoparticle has a cross-sectional area a million times greater than the dye molecule fluorescein, making noble metal nanoparticles exhibiting SPR very attractive materials for ultra-sensitive sensors [4]. There have been many excellent tutorial papers and reviews on SPR that the reader is directed to for more in-depth explanation, as well as the many applications, from sensing to *in vivo* and *in vitro* biomedical applications [2–5].

2.3.2 Optical: quantum dots fluorescence

Metals conduct electricity because their mobile electrons can move through the material. Electrons are confined to their atomic orbitals in an insulator material. So what is it that allows electrons to move in metals and not in insulators? The answer is there are simply vacant orbitals (holes) to move into at the same (or very similar) energy of the mobile electrons in metals. This is described by band theory, where there is an energy band full of electrons, so there are no holes available for electron movement. This is the valance band. Above this is the conduction band. Metals have electrons in the conduction band, which is not full, so there are plenty of vacant sites to move between. Semiconducting materials have a full valance band and no electrons in the conduction band (similar to an insulator), but they differ from an insulator as they have a very small band gap between the two bands. As such they become conducting when the temperature is increased sufficiently for the electrons to be promoted into the higher energy conducting band, leaving an electron hole in the valence band (with both mobile electron and holes contributing to conduction). Examples of semiconducting materials include CdSe, InAs and GaP. Describing electronic properties with respect to bands is only possible in the bulk phase (as bands are a continuous macroscale phenomenon). As the particle size reduces in size to the nanoscale, the electrons become spacially confined (with fewer vacant orbitals to move into) and the bulk models break down. When the size of the particle becomes comparable to or smaller than the electron/hole pair (exciton) Bohr radius (usually $\leqslant 10$ nm), the electrons are confined in all directions and the material is considered to have zero dimensions (hence a dot); these materials are known as quantum dots or 'artificial atoms'. The spacial confinement results in an increase in the size of the band gap and the banded continuum (valance and conducting bands) breaking into a quantised energy level, perfectly demonstrating how the nanoscale is the intermediate between the bulk and the atomic scale. When energy is supplied to a quantum dot (in the form of electromagnetic radiation), an exciton is created as the electron is promoted into the conduction band, and this then emits fluorescence when the electron relaxes back down into the valance band, recombining with the hole. The wavelength of this fluorescence is dependent on the size of the band gap ($\Delta E(r)$) which is dependent on the band gap of the bulk material (E_{gap}), plus an increase in this size as the radius of the quantum dot (*r*) decreases (this relationship is shown in equation (2.4) and figure 2.6). The higher energy (shorter wavelength/ higher frequency) emission is generated from relaxation across a larger band gap which occurs for smaller quantum dots. Equation (2.4) can be rearranged to equation (2.5) to show the relationship of the quantum dot size (radius *r*) to the change in band gap energy (*h* is Planck's constant, m_e is the mass of a free electron and m_h is the mass of the hole).

$$\Delta E_{nano}(r) = \Delta E_{bulk} + \left[\frac{h^2}{8r^2} \left(\frac{1}{m_e^*} + \frac{1}{m_h^*} \right) \right] \quad (2.4)$$

$$r = \sqrt{\left(\frac{h^2}{8(\Delta E_{nano} - \Delta E_{bulk})} \right) \left(\frac{1}{m_e^*} + \frac{1}{m_h^*} \right)} \quad (2.5)$$

Quantum dots have exceptional and far superior fluorescent properties in comparison to traditional organic dye molecules. (1) Due to the relationship between size and emission their wavelength can be precisely tuned. (2) They can absorb a broad range of energy but have a very narrow emission band. (3) Their fluorescence is unparalleled with respect to brightness, lifetime and resistance to photobleaching, which makes them excellent optical probes. One downside is many of the best semiconductor materials for high florescence quantum dots with the band gap at the best wavelength for visualisation are often the most toxic (such as CdSe and Cd/S). This is being addressed by coating these materials in less toxic semiconductors, such as ZnS, and also developing new quantum dot materials by doping less toxic semiconductors to tune the band gap. Again, different morphologies can offer

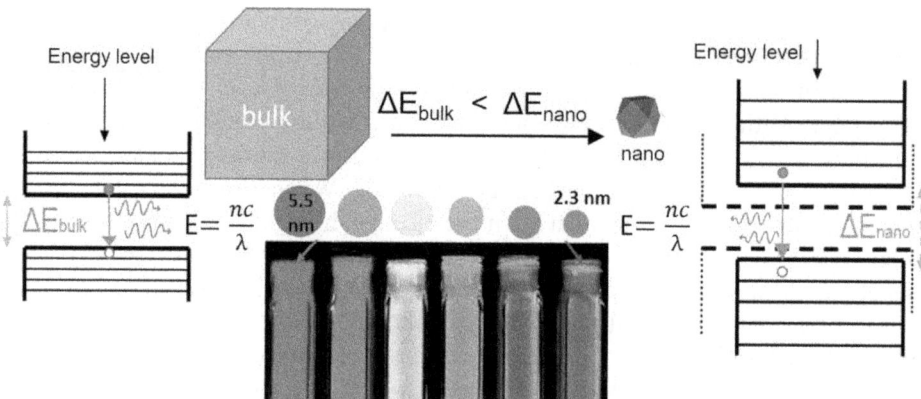

Figure 2.6. Description of how a band gap in a semiconductor increases as the size of the particle decreases. The left-hand side shows the band gap in the bulk material, while on the right an increased band gap is shown for the nanoparticle (bulk gap is shown as a dashed line on the right for comparison) as the frequency of the light emitted is proportional to this energy (and dependent on the material), the light emitted varies with size. CdSe/ZnS quantum dots of decreasing sizes from left to right are shown in the centre (image from [2] reproduced by permission of the Royal Society of Chemistry).

further tuning. Further reading on quantum dots and particularly their biomedical uses and the implications of their toxicity can be found in [6].

2.3.3 Electron spin and nanomagnetism

Nanomagnetism is another perfect illustration of a property on the boundary between bulk and atomic scale. Again we will see here how some nanoparticles behave like atoms (superparamagnets) while others behave more like the bulk (single-domain nanomagnets). Furthermore, we see how the quantum interplay between charge, electron spin and magnetic field can be harnessed in nanomaterials for spintronic applications. To understand the nanoscale properties we must first understand the interplay of electrons and magnetism at the atomic scale, appreciate how this builds to bulk scale, then interrogate the middle.

In fundamental terms, a magnetic field is generated when there is movement of charge. This is seen by the magnetic field surrounding a wire conducting electricity, which is due to the charged electrons moving in the wire (figure 2.7(Ai)). Electrons in an atom have two sorts of motion (orbital and spin (figure 2.7(Aii))) and as such have a magnetic moment: generally it is the electrons in a material that give it its magnetic properties. The moment of a single electron can be calculated from first principles, and is found to be

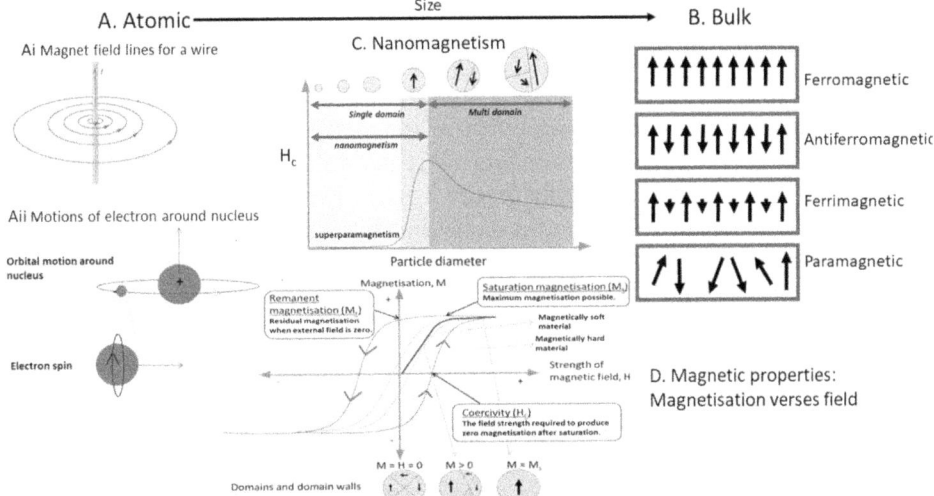

Figure 2.7. Description of magnetism from the atomic to the bulk. (A) Two demonstrations of magnetic fields generated by electrons. (Ai) Macroscale example of magnetic field lines of magnetic field generated by a wire conducting electricity due to the flow of electrons (the current I) in the wire. (Aii) The motion of an electron within an atom (both orbital and spin) which generates an atomic magnetic field. (B) Any unpaired electrons contribute to a bulk magnetic property depending how they are arranged in the solid. Each arrow represents the direction and magnitude of a paramagnetic atom. (C) The nanomagnetic properties, demonstrating how the largest single domain magnetism have the highest coercivity due to no loss of energy through domain wall formation, while superparamagnetic nanoparticles have near zero coercivity. (D) Description of the magnetic properties of hysteresis, plotting the magnetisation with increasing field, then reversing the field and increasing again.

$$\mu_B = \pi r^2 I = \frac{e\hbar}{2m_e} = 9.274 \times 10^{-24} \text{ A m}^2, \tag{2.6}$$

where r is the radius of the orbital, I is the current, e is the charge of an electron, m_e is the mass of an electron and $\hbar$ is the reduced Planck constant.

Two electrons only exist in the same atomic orbital if they have opposite spin values (either $+1/2$ or $-1/2$). The electrons are 'paired' in an orbital and these two opposite electronic spins cancel out each other's magnetic moment, so there is said to be no net overall magnetic effect. However, this is not technically true: such orbits actually have a week repulsion to the magnetic field (in the order of a million times less than the unpaired magnetic effect). If an atom has only paired electrons it is **diamagnetic**. The presence of an unpaired electron in an atom gives the atom an overall magnetic moment and such atoms are described as **paramagnetic**. The application of a magnetic field causes these dipoles to align with the direction of an external magnetic field. The fact that unpaired electrons create a magnetic field and in turn are affected by an external field is fundamental to spintronics. This new field uses the quantum physical properties of ultrathin (single atomic thick layer) 2D nanomaterials, and harnesses the effect of fields on electron spin and spin–orbit coupling to convey electronic information rather than the transport of free electrons down a wire. In purely paramagnetic materials the electron dipoles are only weakly coupled, and as a result thermal energy causes these moments to randomly align. Examples of purely paramagnetic materials include many transition metals and rare-earth salts. These materials have a net magnetic moment due to incomplete electron shells.

Bulk magnetism is a cooperative effect of the alignment of many paramagnetic atoms, and as such it is a property reliant on structural order. There are three main categories of bulk magnetic ordering (figure 2.7(B)). **Ferromagnetism** is when all the paramagnetic atoms are aligned in the same direction even when a field is removed, resulting in the material having a magnetic moment. Ferromagnetism is rare, with iron, cobalt, nickel and some rare-earth metals being the only elements to be ferromagnetic. Ferromagnetism is rare as paramagnetic atoms/ions will tend to align in an anti-parallel way; this form of ordering is termed **antiferromagnetic** and will exhibit no net magnetisation as the opposing paramagnetic moments will cancel each other out. Examples include numerous transition metal compounds, such as iron manganese (FeMn). The final bulk magnetic coupling utilises the strong anti-parallel alignment, but results in a net magnetic moment simply by ensuring there is an uneven number of magnetic dipoles in each direction, so the dipoles are not fully cancelled out. A material with this magnetic ordering is called a **ferrimagnet**. Ferrimagnetic materials tend to be made up of different materials or ions, with examples including garnets such as YIG ($Y_3Fe_2(FeO_4)_3$) and ferrites such as the oldest known magnetic material: magnetite (Fe_3O_4).

Magnetic order is dependent on temperature: when the temperature is increased above a certain threshold (blocking temperature (T_B)) (also called Curie temperature (T_c) or Néel temperature (T_N) for ferr(o/i)magnetic and antiferromagnetic materials respectively), specific to that material, the thermal energy destroys the ordering and the bulk magnetism is lost. Above this temperature the material becomes paramagnetic.

It is well known that magnetic materials can become demagnetised over time. This is because a material will gradually form magnetic domains, which are small regions of local dipole alignment within the material, each separated by a domain wall. It is always worth remembering that maintaining order carries an energy cost, so long-range order will inevitably reduce to short-range order (domain), to eventual disorder at any elevated temperature if there is nothing to maintain the order (i.e. a field). Domains form to minimise the material's energy by reducing magnetic poles at the surface to reduce the magnetostatic energy. However, the creation of a domain wall requires energy to overcome this stability, so an equilibrium is reached. If a demagnetised material is placed back in a magnetic field, the domains most closely aligned to the field direction grow as the walls recede at the expense of the misaligned domains, continuing until all the dipoles are aligned with the higher field and the material is said to be magnetically saturated (figure 2.7(D)).

A material that will easily or reluctantly align with a changing magnetic field is termed magnetically soft or hard, respectively, a property that is measured by its magnetic coercivity (figure 2.7(D)). For a soft magnet, simply removing the field (field = 0) is enough to demagnetise the material. However, a hard magnet will retain its moment at zero field, and even in an opposing magnetic field (figure 2.7(D)). The electrons in such materials have a preferred (lower energy) direction of alignment, so more energy is required to realign them. Coercivity can be quantified from a hysteresis plot (figure 2.7(D)) where a material is driven to saturation by application of a cycling magnetic field through both opposing directions. Where there is a hysteresis loop, work is done to oppose the field and this energy is proportional to the area of the hysteresis curve and is dissipated in the form of heat.

If the bulk material's dimensions are reduced to the nanoscale, a critical volume is reached where it costs more energy to create a domain wall than the total magnetostatic energy of the material, and so a single domain magnetic particle is formed (figure 2.7(C)), which is also dependent on the anisotropy of the material. These are the smallest permanent magnets, and have the advantage that energy is not wasted on the formation of domain walls, so coercivity is the highest for this material (by volume). As the size is reduced further the magnetisation energy of the particles approaches the thermal energy ($K_B T$) and so the magnetic dipole of these particles will continually fluctuate with thermal fluctuations. These particles behave like paramagnetic atoms would (not a bulk magnet) and are thus termed **superparamagnetic**. Such particles do not display hysteresis as the particle itself flips in the field. Interestingly there is not a cut-off defined size at which a single domain magnetic nanoparticle starts to display superparamagnetic behaviour. It is entirely dependent on the temperature and the time-scale of an experiment. That is, the same single domain ferromagnetic nanoparticle could be a permanent static magnet in one experiment where the measurement was conducted over a short time-frame or the temperature is low, and superparamagnetic in another experiment when the measurement is taken over a longer time-scale and/or the temperature is raised. More detailed treatment of magnetism can be obtained from excellent textbooks [7].

2.4 Applications

There is now an ever-increasing demand for nanomaterials for numerous applications (figure 2.8) due to their enhanced/unique properties (above), increased surface area, increased reactivities and compatible size for interactions with molecular technologies. Examples range from personal care, such as TiO_2 in sunscreen cream, to audio technology with magnetic nanoparticles in dampening ferrofluids for loud speakers. More sophisticated nanomaterials are now being specifically designed for a wide breadth of challenges, from curing cancer to purifying water: nanomaterials are becoming ubiquitous in every aspect our modern life as we move towards smaller, more compact and more sophisticated devices. There are many applications in areas such as energy (generation, conversion and storage), catalysis and gas sensing. However, this chapter is designed to provide an overview and not to cover the full range of applications. As such, this section will concentrate on the emerging and

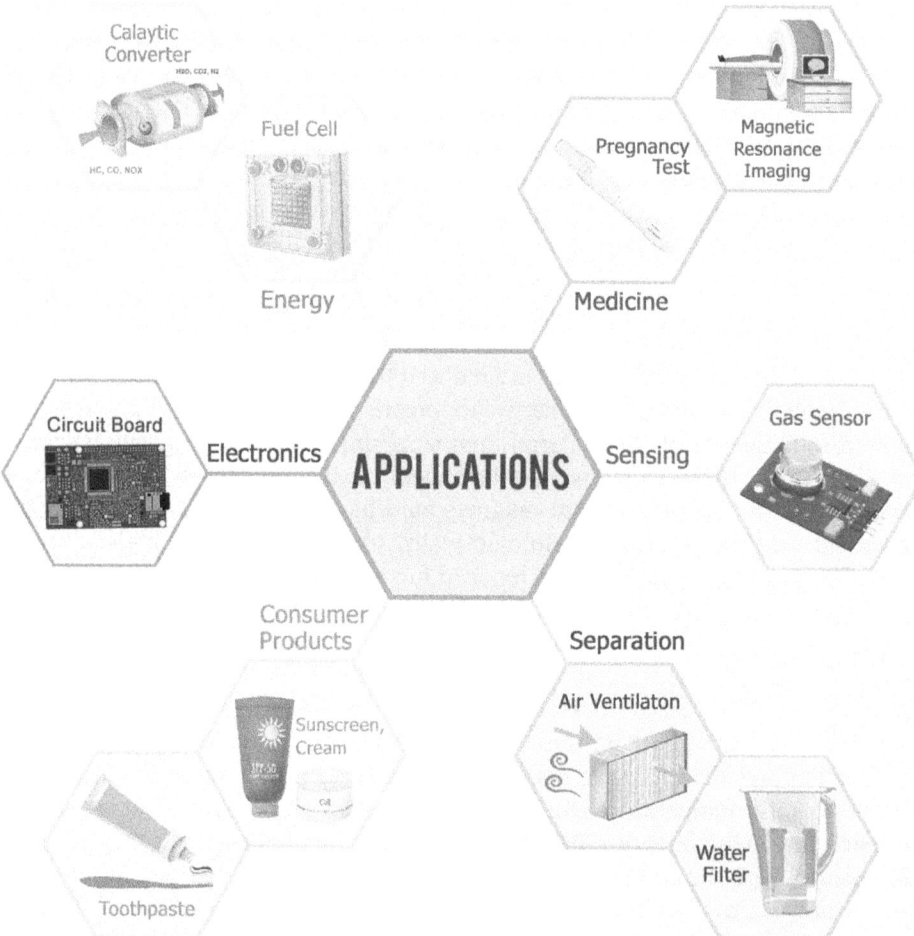

Figure 2.8. Schematic of the types of sectors and applications in which nanomaterials are used.

important field of nanomedicine, as this offers excellent examples that utilise the magnetic and optical properties described in section 2.3. While there is some cross over, we will discuss *in vivo* (in the body) magnetic materials for therapies, with some therapeutic optical materials too. Imaging enhancement material for both *in vivo* and *in vitro* (in a test tube) applications will explore materials with optical properties such as gold and CdSe, and *in vitro* bionano-sensing will be described which takes advantage of surface plasmonic properties. The final two sections will touch on different nanotechnological applications such as data storage and consumer products, to offer some breadth.

2.4.1 Nanomedicine

Healthcare utilises a range of nanomaterials, such as silver in wound-dressing (covered in later sections on consumer products), however, since the 1980s, and with a rapid increased effort in the last 20 years, a new field of biomedical research and application has developed called **nanomedicine.** Nanomedicine is the utilisation of nanotechnology in medicine. This is a really powerful concept, due once again to length scale. As can be seen from figure 2.1, medically relevant biological entities such as proteins and DNA are comparably sized to nanoparticles, making their interaction specific and optimal. Nanomedicine has utilised very specific nanoparticles for imaging disease or specific markers in areas of the body (diagnostics/imaging), or for treatment (therapeutics) or in some cases both (theragnostics) for a range of target diseases. This includes using nanoparticles to highlight specific areas of disease, for either imaging or removal (by magnetic separation), using them as vehicles to deliver drugs to specific disease sites, or using their nanoscale properties such as photothermia and magnetic hyperthermia to treat diseases, with an emphasis on cancer [8]. *In vitro* diagnostic immunoassays will be discussed as the third application, however, the overlapping field of nanomaterials in tissue engineering and regenerative medicine, while fascinating, is too large to topic to discuss here and as such is beyond the scope of this chapter.

2.4.1.1 Generic biomedical nanoparticle
When considering applications *in vivo* there are key factors to consider, the most important being efficacy, toxicity and biocompatibility (the latter two are discussed in section 2.5). These properties along with mode of action can all be controlled via either the particle inorganic core, the coating, or functional biomolecules/small drug molecules that can decorate the exterior. A generic biomedical nanoparticle is shown in figure 2.8(A).

It should be noted that not all of these three components may be strictly necessary (e.g. due to their inert properties, we will we see gold nanoparticles with no coating etc), and the properties above can be delivered by varying components (e.g. the therapeutic can be an active drug molecule on the surface, or heat treatment from the inorganic core). **The inorganic core** offers the nanomaterial's fundamental form. It could be that the particle's key property is simply to be inert and dense (to aid visualisation), but more commonly, the core offers the unique nanoproperties discussed in section 2.3 which can be utilised in the medical application. The main functions of the **coating** are: (1) to provide protection (of the particle against degradation from the environment) and (2) to protect the surroundings from toxic effects of the inorganic material. The two are related: the more stable the particle, the less likely it is to dissolve and prove toxic to the body by leakage. The coating is vital for nanomaterials with unstable or toxic cores such as CdSe quantum dots. These are regularly coated with ZnS and then again with an organic coating to prevent toxicity and increase biocompatibility. The second main function is to aid the particle's biocompatibility. Popular coatings such as dextran are cheap and increase biocompatibility. It is well known that coating with polyethylene glycol (PEG) increases blood circulation times by preventing nanoparticle removal by the immune system. The uptake of modified nanoparticles by cancer cells can be increased by coating in a pH-sensitive zwitterionic coating, that becomes charged at the cancer site (lower pH), so is more readily uptaken, increasing the accumulation at the cancer site [9]. Using biologically derived coatings provides further 'stealth' by invading the immune system and targeting cancer sites by disguising the nanoparticle as the immune system, using macrophage membranes as coatings [10]. The final function of the coating is to enable easy attachment of the third component: active biomolecules. This can be achieved through a range of chemistry, mentioned briefly in later sections. Finally, the nanoparticle can be decorated with **active biomolecules or small molecular drugs** that can be used to target the nanoparticle to the site of the disease, or can be a drug or biological therapies to treat a disease (discussed in more detail in the therapeutics section).

This field is far-reaching, so to keep the content of this section focused on the types of nanoparticles discussed in later chapters and the properties we have discussed earlier in this chapter, only imaging/diagnostics and therapeutics will be discussed here. This critical area of nanomedicine utilising nanoparticles is still in its infancy, with only a handful of nanomaterial-related nanomedicines being FDA approved so far [11].

2.4.1.2 Imaging
There is a vast range of medical imaging techniques with various purposes and advantages. Many (e.g. ultrasound scanning) are not enhanced by inorganic nanomaterials, so will not be discussed further in this section. Some (such as positron emission tomography (PET)) are starting to benefit from enhancement with nanoparticles, but expand into the science of radioisotopes, which again is beyond the scope of this section (see [20] for the development of PET with nanoparticles). Nanomaterials can act as a visualisation probe (highlighting an area of interest) in many microscopy and tomography techniques (see chapter 3 for techniques), or enhance the signal for a specific area, because of its nanoproperties such as nanomagnetism in magnetic resonance imaging (MRI) and magnetic particle imaging (MPI). The nanomaterials become diagnostic probes by being decorated with biomarkers such as antibodies that will target individual diseases, thus specifically detecting and emphasising these areas.

Probes for microscopy can be used both *in vitro* and *in vivo*. Gold nanoparticles have been used extensively on both accounts. Gold can be decorated effectively (e.g. with antibodies) by taking advantage of the strong gold–sulphur bond using a cysteine residue which contains a sulphur. Thus, gold nanoparticles coated with cancer-targeting antibodies (e.g. anti-EGFR (epidermal growth factor receptor)) can be used to visualise cancerous cells with optical microscopy *in vitro* (figure 2.9(Bi)) [12]. Similar optical microscopy can be used to visualise targeted quantum dots using confocal microscopy [6]. The density of gold nanoparticles can also be utilised as a probe for TEM imaging in what is commonly referred to as immune gold staining, where gold nanoparticles can be targeted to a specific biological site. TEM analysis will then identify structures decorated with dense gold dots, distinguishing them from the other biological tissue (figure 2.9(Bii)). Recently, a combined approach of both gold and CdSe quantum dots have been used to image a cell, highlighting different aspects (the gold to the nucleus, and the quantum dot to the insulin granules) offering multimodal enhanced imaging (figures 2.9(Biii), (Biv)) [14]. *In vitro* imaging applications are considered mainly to conduct scientific research and cross over into the area of nanosensing (section 2.4.1.5). *In vivo* imaging is where the real medical applicability lies. Again, gold nanoparticles have been used in diagnostic imaging, using their SPR properties in optical coherent tomography (OCT) [15]. This non-invasive, high resolution optical technique relies on reflective interference between emitted and backscatter light (akin to an optical ultrasound), so can penetrate biological tissue making it applicable *in vivo*. Gold nanoparticles of various morphologies (tuned so the SPR lies within the emitted light source) make ideal contrast agents for OCT as the gold nanoparticles absorb this specific light wavelength, enhancing the image at this location. Hollow nanocages have proved to be the best contrast enhancing agent, as they offer the strongest absorption in the near infrared. Figure 2.9(Bv) shows a phantom with and without enhancement with gold 30 nm nanocages [15].

Magnetic resonance imaging (MRI) is a non-invasive technique which has been used internationally since the 1970s. The technique makes use of the massive number of protons in the body. In basic terms, when a magnetic field is applied to the body (in an MRI scanner) the net magnetisation of protons in the body aligns with the direction of the field. A radio frequency is then applied giving the protons energy. When the radio frequency is removed the protons then relax back, and this relaxation can be measured. Using gradient magnets of varying strength allows the targeting of specific areas for imaging of all three dimensions. Magnetic nanoparticles have great potential as MRI contrast agents, as the interaction of the particle's local magnetic field with the surrounding protons can significantly enhance contrast *in vivo*, by increasing relaxation time, resulting in the darkening of this area of the image, without the toxicity of gadolinium-based agents. The size of the MNPs greatly affects the MRI signal, offering tunable contrast modalities [21].

Magnetic particle imaging (MPI) is a new non-invasive imaging technique that images the magnetic signal from superparamagnetic iron-oxide nanoparticles (SPIONs). Utilising the inherent magnetic properties (from the electrons) results in far greater (×22 million) intensity compared to MRI (which utilises the nuclear

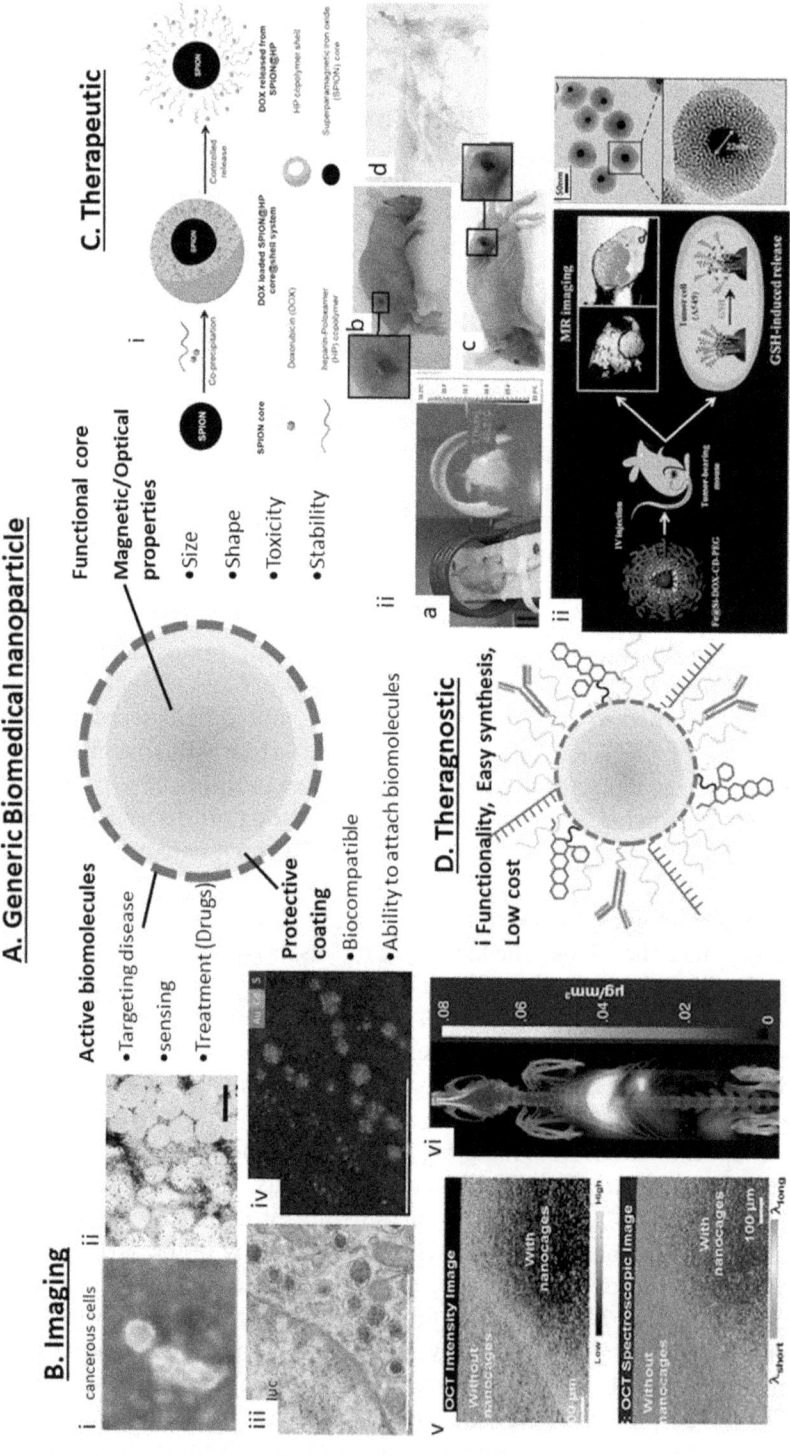

Figure 2.9. Summary of some inorganic nanoparticle uses in nanomedicine. (A) Schematic of and key attributes of a generic biomedical nanoparticle. (B) Some examples of how nanoparticles have been used for imaging. (i) Light scattering image of malignant epithelial cells labelled with antibody (anti-EGFR) coated gold nanoparticles (reproduced with permission from [12], copyright 2005 American Chemical Society). (ii) TEM image of human deep posterior lingual gland selectively labeled with immunogold nanoparticles. Scale bar 1 μm (reproduced from [13], copyright 2006 with permission of Elsevier). (iii) and (iv) Targeted labelling of the nucleus of a cell with gold nanoparticles (red) and insulin granules with quantum dots (green/blue) (scale bar 2 μm) (reproduced from [14]). (v) Spectroscopic optical coherence tomography (SOCT) images of a phantom (top intensity and bottom spectroscopic image) showing contrast due to nanocages (reproduced with permission from [15], copyright The Optical Society). (vi) MPI of a mouse with a lower right flank tumour 9 h after injection with 5 mg kg^{-1} of (superparamagnetic iron-oxide nanoparticle) SPION tracer. A CT scan of the skeleton is overlaid (reproduced with permission from [16], copyright 2017 American Chemical Society) and (C) shows some examples of how nanoparticles are being researched for therapeutic use. (i) Schematic showing the formation of the DOX-loaded SPION@HP core@shell system, and the sustained release at a steady rate of DOX by diffusing through the polymeric matrix as the HP shell degraded. Reproduced from [17]. (ii) An *in vivo* experiment of the treatment of a tumour with magnetosomes and a SPION control. (a) Shows the experimental setup in the alternating field coil, (b) shows the shrinkages of the tumour with treatment of magnetosome chains compared to (c) treatment with SPIONs, and (d) shows the cancer cell with magnetosome chains after 24 h (reproduced from [18] with permission from Taylor & Francis Ltd). (Di) Schematic of a multifunctional combined imaging and therapeutic nanoparticle termed a theragnostic nanoparticle. (ii) Schematic representing the dual function of Fe@Si–DOX–CD–PEG nanoparticles having both MRI contrast agents and for SMART targeted drug release. Fe@Si cores are magnetite nanoparticles coated with nanoporous silica shown in the TEM image on the right. DOX is the anticancer drug doxorubicin located in the silica pores. CD is cyclodextrin which can release the DOX when exposed to glutathione (GSH) which is present in high concentrations in cancer cells (adapted from [19]).

moment). SPIONs are targeted to a site in the body then imaged by applying a saturating magnetic field, then rostering an excitation field, which causes the SPIONs to flip, which can be detected at this location in 3D. It has been used to successfully locate and image cancer tumours (figure 2.9(Bvi)) [16] but its sensitivity means it could be used for a range of more subtle diagnostics too.

2.4.1.3 Therapeutics
Targeted drug delivery uses a nanoparticle (for this book, specifically an inorganic nanoparticle) as a vehicle to transport a drug to a specific disease cell sites. Targeted drug delivery is a game-changer for global pharmacology. Most cancer drugs' activity depends on their toxicity to cancer cells, but the obvious issue is that these are invariably toxic to healthy cells too. This means that previous therapies had terrible side effects, as the body had to cope with being poisoned overall in order to poison the cancer tumour. Targeted drug delivery means that the drugs are delivered to just the cancer cells, and thus are not administered to the whole body. This has two advantages: (1) the side effects of the treatment are dramatically reduced as the effect on healthy cells is now negligible; (2) the quantity of drug required is vastly reduced, as it is only used on the much smaller affected area. Targeting can be achieved passively or actively.

Passive targeting takes advantage of the fact that tumours have more 'leaky' vasculature and faulty drainage, so have enhanced permeability and retention (EPR). This means that tumours are susceptible to nanoparticles of between 12–500 nm accumulating preferentially within tumours, without any other 'active targeting'. However, accumulation quantities can be significantly increased by active targeting also. Active targeting can take the form of specific antibodies, as discussed above, which will target just the specific cancer cells. There are also other physiochemical properties that differentiate cancer cells from healthy cells which can be taken advantage of, such as pH. Cancer cells are more acidic, and as such a pH-responsive nanoparticle coating can be designed to release the drug only at the pH of the cancer cell, targeting the drug release.

Magnetic nanoparticles can be further exploited for targeting by utilising a magnetic field gradient to move the particles remotely *in vivo*. Magnetite or maghemite SPIONs have been most widely used due to their biocompatibility and easy availability for iron precursors. A permanent magnetic field near the tumour can enable the field gradient to concentrate and retain the drug carrying magnetic nanoparticles at the tumour site. However, depending where the tumour is, this will be more difficult when in deep tissue or internal organs. In these cases the field gradient in an MRI instrument can be used to concentrate the magnetic particles in all three dimensions at any location in the body [22].

The most studied SPION–drug conjugate is those of the intercalating chemotherapy drug doxorubicin (DOX). Doxorubicin is a well know and characterised potent chemotherapy agent, but it also has two further advantages for drug delivery design and research. The first is that it has two functional groups that can be modified to conjugate to the drug delivery vehicle, and the second is the conjugated aromatic nature of the drug makes it fluorescent, so it can be easily tracked and monitored. Researchers have designed a range of SPION–DOX nanomedicines ranging from the simple to the complex. For example, simply trapping SPIONs and unmodified DOX in polymer micelles/vesicles, or trapping unmodified DOX within the polymer coating on a SPION. Such a polymer coating can be specifically designed to change conformation or degrade to release the DOX on under the influence of a specific stimulus, or gradually over time for controlled slow release (figure 2.9(ci)) [17].

Examples of more complex systems include attachment of the DOX to the SPION through either iron coordination chemistry [23] or by conjugation to a responsive synthetic or biological polymer such as a tumour-targeting aptamer [24]. More complex systems include multiple nanoparticles such as SPION with quantum dots, gold nanoparticles and nanoporous silica for imaging and drug entrapment respectively, forming theragnostics, described in section 2.4.1.4. The details above have provided a brief sample of the extensive research being performed, to give an idea of the developments of drug delivery using nanoparticles. This field is developing all the time, and as such the information above will date rapidly. For more in-depth reviews [11, 25] of this subject, please see recent reviews and up-to-date literature.

Magnetic hyperthermia therapies are emerging non-invasive therapies, currently undergoing clinical trials, which work by causing magnetically-induced heating when magnetic nanoparticles are subjected to an alternating magnetic field (at frequencies in the range of 100–150 kHz). This heating acts to ablate and kill the diseased tissue at a higher localised temperature (above 45 °C) while lower temperatures between 40 °C–44 °C will sensitise the cell making them more susceptible to chemotherapy drugs. Lower temperatures still (39 °C–42 °C) could also be used to trigger a heat-activated drug or release a drug conjugated through a heat-activated linker [26, 27].

The heating is induced by one of two mechanisms. The alternating field can either switch the individual spins of larger single-domain MNPs, or physically flip the whole particle for smaller superparamagnetic MNPs. This switching/flipping induces heating as thermal energy which is released as the spins/particles attempt to resist the alternating field. This heat dissipates into the surrounding tissue, damaging or killing the diseased tissue or activating smart drug treatments. The first mechanism is called ferromagnetic hyperthermia, and describes the thermal energy generated by the switching dipole moments in the crystal lattice by Néel rotations. In effect the heating can be described by the area of the magnetic hysteresis loop (see equation (2.7)) so is increased with the increased magnetic saturation and coercivity of the MNPs. The second mechanism is called magnetic fluid hyperthermia, and is described by the Brownian rotation when the aligning of the magnetic moment causes the whole particle to rotate in superparamagnetic MNPs, causing shear stress as the particle flips resulting in delivery of thermal energy to the solution, which can be described by equation (2.8).

$$P_{FM} = \mu_0 f \oint H \, dM \tag{2.7}$$

$$P_{\text{SPM}} = \mu_0 \pi f \chi'' H^2 \qquad (2.8)$$

where P is the power of the heat generated, μ_0 is the permeability of free space = $2.566\,3706 \times 10^{-6}$ m kg s^{-2} A^{-2}, f is the frequency and H is the magnitude of the alternative magnetic field. The integral of $H\,\mathrm{d}M$ is the area of the hysteresis curve, while χ'' is the out-of-phase component of the magnetic susceptibility of the SPM MNP (effectively how much M lags behind H at that frequency). Within each regime the heating is largely dependent on the magnetic saturation/susceptibility and field frequency. While there is coercivity dependence in single domain ferromagnetic hyperthermia, the effect of the magnitude of the field is squared in magnetic fluid hyperthermia, so can lead to much greater heating. Furthermore, it should be noted that both mechanisms can operate in the same sample where the particles are small enough to rotate and have some coercivity, giving them both Néel and Brownian relaxation [27].

The heating power of a particle is determined by calculation of its specific absorption rate (equation (1.1)), which calculates the heating power (P) of the particles based on the mass of the MNPs in the sample (m_{mnp}). Each SARs value is for a specific field and frequency, so the intrinsic loss parameter (ILP) is the term normalised for field and frequency, both of which are shown in equations (2.9) and (2.10) respectively:

$$\text{SAR} = \frac{\Delta T}{\Delta t} \frac{c}{M_{\text{Fe}}} \qquad (2.9)$$

$$\text{ILP} = \frac{\text{SAR}}{H^2 f}. \qquad (2.10)$$

As with MRI and MPI the potential for MNPs in nanomedicine is far reaching, but particle size along with field and frequency must be considered, as all these factors will affect the heating power. For example, at the same frequency of 100 kHz, 8 nm iron-oxide MNPs gave the best increase in temperature of 9.3 °C in lower fields (9.6 kA m^{-1}), whereas at higher fields (23.9 kA m^{-1}) the 8 nm particles showed an improved increase in temperature of 25 °C, but MNPs of 24 nm in diameter proved more effective with increases in temperature of 55 °C. This is probably due to the combined effect of both heating mechanisms for these larger particles [28].

There are numerous examples in the literature of MNPs used to treat cancer cells from *in vitro* tissue culture to mouse model studies, both as independent therapeutics and in combination with drugs that are improved when cancer cells are heat sensitized [29]. SPM nanoparticles within polymersomes have been extensively probed for this purpose and ILP in the range of values between 1.6 and 5 nH m^2 kg^{-1} have been extensively shown to be functional for hyperthermic therapies [30]. Similarly, bacterial magnetosomes have shown excellent promise for magnetic hyperthermia with the highest recorded ILP of 23.4 nH m^2 kg^{-1} [31, 32], and with vastly increased SARs values compared to comparatively sized synthetic magnetite nanoparticles at lower fields [33]. It is not clear why these biomineralised MNP have such enhanced properties, but is has been speculated that it could be due to the monodispersity or the slightly reduced nature of the magnetite core. Magnetosomes have thus been investigated in mouse models *in vivo* and it has been found that magnetic hyperthermia using chains of magnetosomes has a much greater impact for tumour shrinkage than SPION control particles (figure 2.8(Dii)) [18].

Again, the field moves fast, so recent reviews [29] and research papers in the literature are the best sources of information.

Photothermia is another type of localised hyperthermia that takes advantage of the energy a metallic nanoparticle receives from an electromagnetic wave (see section 2.3.1), where this absorbed photon energy is transferred to phonons in the lattice and dissipated as heat. A relatively new phenomenon, the potential of photothermal therapies has only been realised over the last 15 years. As such it has not yet been subjected to the same level of theoretical treatment as magnet hyperthermia. The electromagnetic wave is provided by a near-infra-red laser (typically an 808 nm laser powered at 0.3–5 W cm^{-2}), which crucially is more transparent to body tissue and can penetrate deeper into the tissue than optical light. While the science in this area is still at the research stage, several inorganic nanomaterials are proving to be ideal materials for photothermal therapies. The front runner is gold nanoparticles (see recent reviews [8, 34]), as well as carbon nanotubes (CNTs) and MNPs such as SPION and larger magnetite MNPs. Once again, bacterial magnetosomes have proved to be excellent candidates with increases of temperature of up to 50 °C (under 1 W cm^{-2} laser power) [31]. Meanwhile, synthetically produced magnetite MNPs coated in macrophage biological coating have proved to be excellent photothermal agents. The macrophage coating renders them invisible to the immune system, offering them longer circulation time and also targeting them to the cancer cells. Furthermore, the magnetic properties of the particles mean they can be magnetically targeted to the tumour *in vivo*, with photothermal therapy resulting in tumours five times smaller by weight compared to treatment with non-macrophage coated MNPs [10].

2.4.1.4 Theragnostics

Theragnostics is quite simply the combination of both an imaging/diagnostic modality with a treatment of the disease (figure 2.8(Di)). It stands to reason that once a nanoparticle has been designed to locate a disease and visualised it, it would seem sensible and economical to provide a drug-payload or therapeutic treatment on the same particle, which has already reached its target. There is some debate as to whether or not theragnostics falls between two stools, in the sense that designing a combined 'all singing, all dancing' nanoparticle may end up doing both jobs to an okay standard, whereas actually designing a particle for one specific purpose may achieve better results. However, as we have seen throughout this section on nanomedicine, many nanomaterials, as well as coating and drugs, are multifunctional, so many theragnostic treatments are thus widely reported in the literature. For example, MNPs can use magnetic force to target the MNP to the site of a tumour, can be visualised with MRI and MPI, and can be heated with either a magnetic field or photothermia. Gold can be visualised with SPR and heated with photothermia, while both can be decorated with drugs and targeting agents such as biomarkers and antibodies and coated with biocompatible/targeting/smart activating polymers/membranes. Thus, a therapy can quite often couple as a imaging tool/diagnostic without any extra complexity and thus biological or synthetic penalty, and this is where the real advantages lies. The number of theragnostics currently being researched is too vast to describe here, so we will examine one example with multiple features. Figure 2.8(Dii) shows Fe@Si–DOX–CD–PEG [19]. Here a SPION forms a magnetic centre which can be magnetically targeted to the tumour. It can also be used as an MRI contrast enhancer offering targeting and imaging capability. The SPION is then coated in nanoporous silica, in which DOX can be loaded into the pores acting as a drug trap. This particle is then coated with PEG to provide biocompatible and cancer responsive cyclodextrin, which blocks the pores until the nanoparticle reaches the cancer site. At the site the cyclodextrin is cleaved off releasing the DOX from the silica pores [19].

2.4.1.5 Nanosensors in healthcare and medicine
Nanomaterials are ideal sensors because of their very large surface area. Furthermore, because of their small size, they are comparable in scale to many biological targets/disease markers making them particularly useful in biological sensing. Finally, the specific nanoscale properties covered in section 2.2 (particularly optical SPR) are very dependent on subtle chances at the surface of the particles, yielding very sensitive sensors.

Within medical sensing, the central dogma for the last 50 years has been the use of antibodies raised against a specific biological target to detect that target. Initially in the 1960s this was done with radio labelling of the antibodies and detecting radioactivity. While this is a very sensitive technique, it is not very accessible and has many safety drawbacks. The 1970s saw the development of more accessible immunosorbent techniques such as the enzyme-linked immunosorbent assay (ELISA), where detection is achieved through an antibody attached to an enzyme, which when coupled to its substrate produces a detectible response (usually a definitive colour change). There are many forms of this assay but all use an antibody raised against the target to 'capture' the target, and an antibody linked to an enzyme detection system. Different types of ELISA include direct ELISA, where the primary antibody is already functionalised with an enzyme for detection, and sandwich ELISA, which has more steps and uses a primary antibody immobilised on the surface to catch the target and further antibodies to specifically bind to the target. Non-specific antibodies to bind to these which contain the enzyme detection system are then added. One of the best-known examples of ELISA sensing is in healthcare, and is the pregnancy test. This is a sandwich ELISA colorimetric sensor for the pregnancy hormone human chorionic gonadotropin (hCG). The test uses capillary flow to move the analyte (urine) along the test strip and over the immobilised antibodies. Most nanosensors will use antibodies in the form of an immunosorbent assay to target biomarkers and specific proteins, as they are well established, very sensitive, accessible and accurate. However, in recent years there has been a move to use nanoparticles as the detection system instead of enzyme assays, due to potential interactions of the enzyme with the sample and the higher intensity and thus sensitivity of the SPR signal. In this system antibodies can be attached to gold nanoparticles using the strong sulphur–gold bond, by simply adding a cysteine amino acid residue to the antibody. Furthermore, with SPR, there are many more sophisticated improvements that can be made on this simple assay, as this property (outlined in section 2.3.1) is ideally sensitive for nanosensing and can be tuned extensively. Furthermore, much more information, such as spacial information, can be obtained from SPR sensors, which cannot be achieved from simple colorimetric assays.

Increasingly, the SPR light emitted by gold and silver nanoparticles is serving as very sensitive colorimetric detection methods for very small quantities of target. One such target is microRNA (MiRNA). Some of these short strands of RNA are markers for the early signs of some cancers, so would be very useful to detect. However, they are present in very low concentrations. Recent work has used a DNA sandwich assay to detect the specific MiRNA [35]. Here, folded capture DNA containing the complementary sequence to the target MiRNA is immobilised on a gold surface (through a sulphur bond) and unfolds to bind the MiRNA specifically. This subtle change to the surface can be detected by a shift in the SPR signal [35]. Then a gold nanoparticle can be added with more single stranded DNA attached, designed to bind further up the capture DNA only if the MiRNA has bound, and placing the gold nanoparticle in close proximity to the gold surface, greatly enhancing the SPR refractive index signal and changing the resonance angle. This is enhanced further by adding more DNA to create a DNA supersandwich (figure 2.10(A)). This allows detection of MiRNA at concentrations as low as 8×10^{-15} Molar [35].

When two gold nanoparticles are brought close together, the effect of proximity couples their plasmons, red shifting the SPR peak to larger wavelengths as two gold nanoparticles come closer together and blue shift to smaller wavelengths as they move apart. This phenomenon can be used to sense and measure tiny distances and form a time resolved molecular ruler (figure 2.10(B)) [36]! Here the DNA is attached to one nanoparticle, thus a sulphur linker, while the other end is functionalised with biotin which allows it to bind to the streptavidin coated nanoparticle. A shift from green to orange particles is immediately seen on this coupling. The molecular ruler could be used to measure the effect of hybridising the single-stranded DNA (ssDNA) to double-stranded DNA (dsDNA) (figure 2.10(B)) [36].

2.4.2 Nanodevice technologies

2.4.2.1 Lab-on-a-chip diagnostics

There is much overlap between the last section on nanosensors, and this section on nanodevices. Indeed, much of the research in the area of nanodevices is centred around medical sensing and the future goal of completely personalised and rapid multiple diagnostics healthcare built into one small chip, which can then be easily processed in an simple hand-held electronic reader (figure 2.11(A)). Here a drop of blood could be distributed via microfluidics to multiple nanosensors to detect a

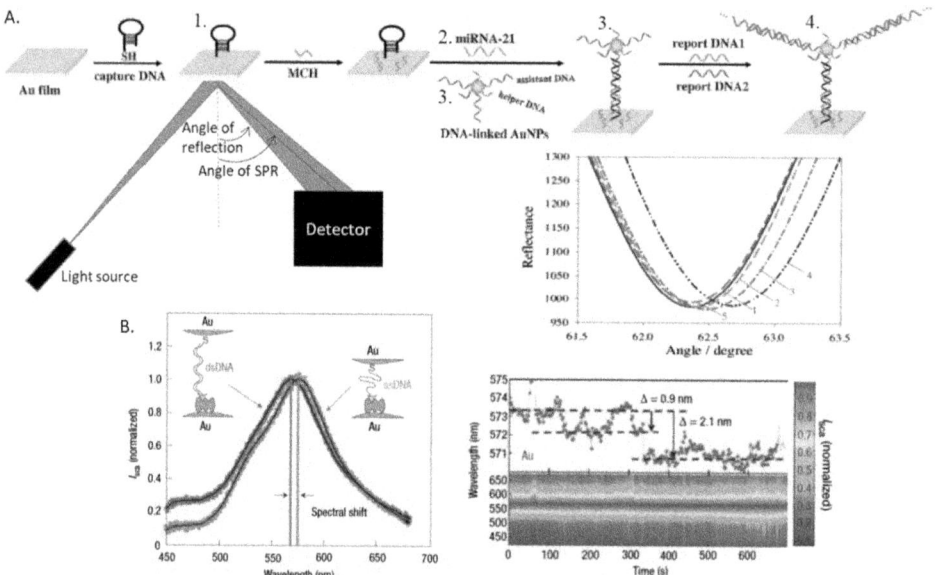

Figure 2.10. Two examples of ultra-sensitive nanosensors using SPR from gold nanoparticles. (A) A DNA supersandwich assay for the detection of very low concentrations of MiRNA. The numbering on the experimental schematic (top) corresponds to the numbering on the data (below right). A schematic showing how the data is collected from the gold surface is shown below left (adapted from images from [35], copyright 2016, with permission from Elsevier). (B) A molecular ruler measuring the extended distance of hybridising flexible ssDNA to more rigid dsDNA. (i) Example spectral shift between a gold particle pair connected with ssDNA (red) and dsDNA (blue). (ii) Spectral position as a function of time after addition of complementary DNA. Discrete states indicated by horizontal dashed lines (reproduced from [36] with permission of Springer).

complete set of diseases and wellbeing factors with unparalleled sensitivity and speed. This idea of bringing multiple nanosensors and microfluidics together on one surface 'chip' has been termed lab-on-a-chip and has coalesced into a dedicated research field driven by this goal, with many excellent reviews on the subject [37, 40].

2.4.2.2 Data storage technologies

In this information age, we see the collection, processing and storage of data and information becoming more important in our daily lives, and also see that the remit for electronic data storage devices goes beyond utility only in a PC. Increasingly consumer products are becoming more user-friendly, intelligent or 'smart', and require the ability to access and store information. The invention of the magnetic hard disk drive (HDD) in 1956 by IBM was the birth of modern information storage as we know it, and magnetic HDDs are still found within most modern PCs (figure 2.11(Bi)). They remain the most commercially viable choice for high volume data storage, something that may become more important if the current shift towards cloud computing continues.

A magnetic HDD uses the same principle of magnetic tape recording, where an electronic signal writes information in a magnetic pattern on the tape. For a

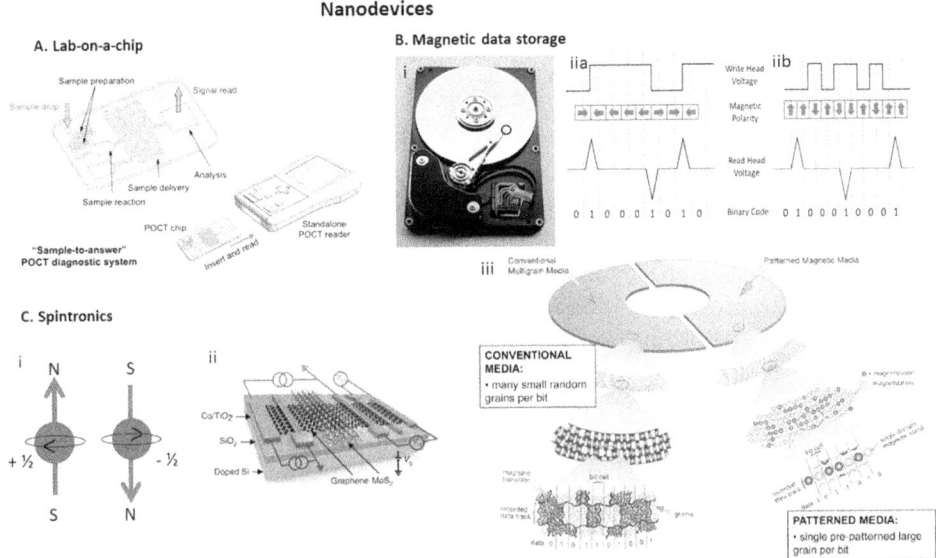

Figure 2.11. (A) Diagram of a point-of-care test chip and reader to offer immediate diagnostics (reproduced from [37], copyright 2015 with permission from Elsevier). (B(i)) Image of a hard disc drive (HDD), (ii) diagram describing how a magnetic patter can be read as binary code for (a) transverse magnetic recording and (b) perpendicular magnetic recording (reproduced with permission from Scott Bird). (iii) Schematic of conventional granular magnetic recording media (left) compared to high-density patterned media (right) (reproduced with permission from [38], copyright 2008 IEEE). (C(i)) Schematic to show the central premise of spintronics. The electronic spin can be manipulated with a magnetic field. (ii) A 2D spin field-effect switch of MoS_2 on graphene (reproduced with permission from Macmillan Publishers Ltd [39], copyright 2016).

magnetic HDD, this information is recorded in binary from 'bits'. Simply, the two binary components synonymous to either of two magnetic spins align in the same direction (0) as the neighbour, or the opposing direction (1) in longitudinal magnetic recording (figure 2.11(Biia)), or align anti-parallel (0) or parallel (1) to their neighbour in perpendicular magnetic recording (figure 2.11(Biib)). The other key difference between magic tape recording and a magnetic HDD is that an HDD can be randomly accessed rather than being read and written serially. This is possible due to the nature of the magnetic medium being a granular thin film on a spinning disc which can be accessed anywhere by a flying read–write head (figure 2.11(Biii)). The data storage capacity and economy of HDDs have vastly increased over the years: 2011 saw storage capacities of 500 Gb in^{-2} at a cost of \$0.2 per GB compared to 2 kb in^{-2} in 1956 for the first 305 Random Access Method of Accounting and Control (RAMAC), which weighed over a ton and cost \$3200 per month (equivalent to approximately \$160 000 today) to lease. The revolutionary nature of computing and electronic data processing/storage was not imagined 60 years ago, when Thomas Watson (then chairman and CEO of IBM) famously said 'I think there is a world market for maybe five computers'. While the development over time has been phenomenal, there is still ever-increasing demand for miniaturisation and higher density data storage capacity. However, reducing the size of each bit is reaching its physical limit. It is likely that conventional granular recording mediums will soon reach storage capacities of approximately 1 Tbit in^{-2}, with bit sizes of approximately 625 nm^2 formed by aligning the polarities of multiple grains. At very small sizes the thermal energy is greater than the magnetic stabilisation energy with the onset of superparamagnetism, meaning the magnetic orientation and thus the data is lost at room temperature.

Two key new approaches to magnetic data storage are currently in development, with the aim of continuing the trend of increasing storage densities and capacities. These are: energy assisted recording and lithography patterned media (figure 2.11(Biii)). Energy assisted recording applies energy in the form of heat (heat-assisted magnetic recording) or microwaves (microwave-assisted magnetic recording) to the granular recording media, allowing the write head to orient the magnetic polarity of a higher anisotropy medium. Lithographically patterned media replaces the featureless granular recording medium used in current magnetic HDDs with a nanopatterned magnetic medium [38]. In nanopatterned media the size of the bit is defined by the nanopattern and not the write head, with the non-magnetic spaces reducing the risk of magnetic coupling and disorientation between neighbours, overall increasing the areal density. These patterns can take the form of discrete tracks (discreet track media) or each bit being retained on a discreet nano-island or nanoparticle (bit patterned media (BPM)) (figure 2.11(Biii)) [41]. With the use of current head materials thermally stable magnetic islands or nanoparticles suitable for BPM could be as small as 8 nm in diameter; with the addition of spaces between the islands' bit sizes could be as small as 12 nm^2. As a result it has been predicted that storage densities could be pushed beyond 50 Tbit in^{-2}. However, the development of a cost effective and industrially scalable manufacturing technique for BPM mean that currently this technology is still very much in the development stage [41].

Although common in computer systems, magnetic hard disk drives (HDDs) are not the only technologies available, with optical and solid-state semiconductor drives now found in many products. Semiconductor memories are electronic data storage devices which are compact and can operate at very high speeds due to a lack of moving parts. As such these devices have become the primary internal memories within computers. Semiconductor memories are either termed 'volatile devices' which lose stored information without a power source, or 'non-volatile devices' which will retain stored information in the absence of power. The two most common semiconductor memories, which fall into the volatile class, are dynamic random access memory (DRAM) and static random access memory (SRAM) [37]. DRAM stores bits of information inside capacitors, with each capacitor encoding a bit of information. A 1 or 0 can be formed by the capacitor being either charged or discharged, but these capacitors will discharge if the charge is not continually refreshed. Hence information stored on these devices will be lost if power is lost also. The recent trend towards portable computers, smartphones, as well as portable electronics has fuelled the development of fast and compact non-volatile semiconductor data storage [42]. One of the most successful technologies in this category is flash memory, which can store data for approximately ten years without power. Flash stores data in arrays of transistors, which make up a memory cell. As with magnetic data storage, flash memory has seen massive improvements in storage densities and capacities by scaling the size of the transistors to smaller dimensions. Again, this process cannot continue indefinitely; with the physical limits being reached, new alternative technologies that utilise the quantum physical properties of nanomaterials are being explored [42].

2.4.2.3 Nanoelectronics/spintronics

As electronically powered technology has now become vital to our lives, we need more and more efficient methods of transferring electrical energy. Furthermore, the reduction of the size of electronic technology needs consideration. In macroscale electronics electrical energy is transferred by electrons in a metal wire moving through the lattice to form a current; charge difference can build up between two points which have a potential difference between them, which is the voltage. Electronic components have been developed throughout the 20th century, such as capacitors (collecting charge, then releasing it when a critical level is reached), diodes (semiconductors that act as a valve, allowing the electrons to flow easily in one direction but not the other), and transistors (semiconductors that are able to amplify or switch electronic flow), which have allowed our lives to be transformed with ever more complex and ingenious electronic circuity in common everyday devices. However, as size reduces, so too do the physical properties of the material. As wires get thinner, the resistance in the wire vastly increases. As we miniaturise, it may be time to consider whether the movement of the charged electron, i.e. controlling electrical devices by conveying electron charge flowing in a wire, is actually the best method in smaller devices.

Spintronics is a new field of physics studying how the spin of electronics could be a better way to convey and control electronic signal in smaller electronic devices. Apart from the obvious benefit of miniaturisation, this has many advantages. In spintronics, there are additional characteristics of spin, and how the electron spin couples with the atomic orbital (spin–orbit coupling), how electrons can tunnel, and ultimately how these respond to an external magnetic field (figure 2.10(Ci)). All of these can be tuned, offering multiple electronic manipulation methods at the atomic level [43]. As recently as 2010 and 2013, graphite-like structured 2D semiconductor MoS_2 was shown to have photoluminescent and transistor behaviour, respectively. Building monolayered layered structures of such semiconductors, insulated and ferromagnetic materials are paving the way for complex spintronic effects that could be used in the nanoelectronics devices of the future. A 2D layered structure of MoS_2 on graphene combines the strong spin–orbit coupling of MoS_2 with the enhanced spin transport properties of graphene to form a spin field-effect switch capable of both transporting and controlling spin current (figure 2.10(Cii)). Such switches could be used to improve search engines and pattern recognition circuitry capability in the future [39]. However, spintronics is already in technology now: read heads on magnetic hard drives already use the effect of electron tunnelling from ferromagnetic layers through an insulating layer in the material. Combining the set-up of semiconductor RAM devices with magnetic domains for the data storage (instead of electronic charge) and utilising the electronic tunnelling of spintronics has led to the development of magnetoresistive random-access memory (MRAM), which has been heralded to out-compete all other forms of data storage in the future (currently not for high density, but for speed and longevity of retention) [44]. A MRAM is made up of an array of magnetic tunnel junction that holds a bit of information. Each magnetic tunnel junction is composed of a complex layered structure of two ferromagnetically orientated thin films separated by an insulating layer, with the bottom ferromagnetic film 'pinned' in orientation by a further antiferromagnetic thin film below.

The key issue with the development of all the complex nanodevice technologies described above is fabrication. Producing the correct material, of the correct phase and orientation in the correct layered or nanoscale morphological structures at the nanoscale is very problematic. Adding to that the ever-increasing need to fabricate such intricate complex materials and systems in a reproducible and sustainable way, for them to be industrially viable for applications, presents a formidable challenge.

2.4.3 Consumer products

A consumer product inventory has listed in total 1829 products containing nanomaterials, using in total 47 types of nanomaterials (www.nanotechproject.org/cpi) [45]. These products are produced by over 600 companies, which are spread over more than 30 countries globally. Of the nanomaterials used, titania, silica, zinc oxide and silver are the most common both in terms of total mass, as well as the number of products they are found in. When considered on mass basis, titania, zinc oxide and silica are the top three most manufactured nanomaterials.

2.4.3.1 Titania and zinc oxide
Titania and zinc oxide form the active components of sunscreens due to their ability to specifically interact with UV light (absorb or reflect) but not visible light. As discussed in the optical properties section above, it is their size and specific chemistry that enables them to block UV rays (both UV-A and UV-B in the range of 320–400 nm and 290–320 nm, respectively). The interactions of these two materials arise from their band gaps, which allow them to absorb UV rays; the specific wavelengths of absorption depend on the actual band gap (which in turn depends on the chemistry and sizes) of the materials. Generally, titania is most effective in blocking the UV-B range, while ZnO blocks the UV-A range. Using nano-sized (<100 nm) titania or ZnO makes them transparent in visible light, as well as enabling control over their band gaps (which become dependent on the particle sizes, as seen in section 2.2).

2.4.3.2 Silica
Silica is another widely used nanomaterial in applications such as toothpastes, tyres and cosmetics. One of the key properties in its success is its inertness under most conditions of use. Silica is resistant under acidic conditions and most organic reagents—it can only be degraded by strong caustic solutions (pH > 10) or hydrofluoric acid. As silica can withstand high temperatures similar to glass, it

provides excellent improvement in thermal properties when blended with polymers. This, coupled with its added mechanical strength, means silica is an ideal component in tyre and composite applications. Further, silica (in most forms) has been 'generally regarded as safe (GRAS)' by regulating bodies in the USA and Europe, and hence is widely used in food and cosmetics. In those applications, silica improves the rheological (flow) properties of food and cosmetic products. Finally, its specific porous structure and the ability to absorb water/moisture makes it highly useful in desiccant applications where even a small amount of moisture needs to be avoided (e.g. electronics packaging and table salt).

2.4.3.3 Silver nanoparticles
Silver nanoparticles are another major nanomaterial used in consumer products, mainly for antimicrobial protection. The use of particulate silver in healthcare for the treatment of wounds has been recorded as far back in time as Ancient Greece. Silver is a well known antimicrobial agent, being used in antibacterial, anti-fungal and antiseptic creams, dressings and medical equipment coatings. While silver is an inert metal to us, it is highly toxic to microbes by damaging enzymes in pathogen membranes. The advantage of nanoparticulate silver is that the increased surface area increases the relative activity. It is now common to find silver nanoparticles in simple cheap wound dressings such as sticking plasters. The antimicrobial effect of silver nanoparticles is used in a wide range of consumer products, such as coatings and clothing.

The nanomaterials outlined above can also be used in air and water filtration/purification media for domestic use. They can treat air or water by removing pollutants via adsorption (e.g. silica), disinfecting (e.g. silver) or photocatalytic destruction (e.g. titania). They are also used in coatings for optical or mechanical property enhancement (e.g. controlling gloss or providing scratch resistance).

2.5 Nanomaterial biocompatibility and toxicity

Just as the special features of nanomaterials enable new functions and applications as discussed above, the same features can lead to undesirable toxic effects on biological systems and to the environment. Although nanomaterials can cause toxicities of varying nature and degree, results from studies indicate that these effects can be mitigated or controlled by developing a mechanistic understanding of the effects, and hence nanomaterial toxicity is unlikely to become a major problem [46].

Before diving into learning about how nanomaterials can cause toxicity, it is important to point out a key difference between large (not-nano) objects containing nanoscale active areas, and nano-objects where the active area is a discrete nano-particle [47]. An example of the former is a mm-wide electronic chip with nanoscale junctions, where the nano-features do not lead to any added toxicity. An example of the latter case is nanoparticles used in cosmetics or as drug delivery vehicles where individual particles are in contact with the external environment and hence capable of invoking responses due to nanoscale effects. This section only covers such cases.

Biocompatibility refers to efficacy without any adverse effects, such as toxicity. Toxicity of nanomaterials to a life form is typically caused by the ability of nanomaterials to release free radicals or cause oxidative stress when in contact with tissues or cells. These effects are detailed in table 2.1 and they arise from a range of physicochemical properties of nanomaterials. These properties can be categorised in three forms as discussed below.

Chemical composition and functionality: The chemical composition of the core and the surface are important. Let us consider each of these in turn. Comparing only the chemical composition, materials made up of cadmium are highly toxic because it affects cell functions such as proliferation and death by damaging lipids, DNA, proteins and enzymes and potentially causing gene mutations [48]. On the other hand, materials made from silica are generally biocompatible, making it widely applicable in food and cosmetics (see exceptions below) because silicon has been viewed as an element essential to life, including bone and tissue formation in animals and a defence against biotic and abiotic stresses in plants and animals (e.g. toxicity from metals such as aluminium) [49].

Surface chemistry and functionality (in combination with size effects) are very important factors determining the toxicity/biocompatibility of nanomaterials

Table 2.1. Toxic effects of nanomaterials. Note that most of these are supported by limited experimental or clinical evidence. Table adapted from [46].

Effects	Outcomes
Reactive oxygen species generation	Protein, DNA and membrane injury, oxidative stress
Oxidative stress	Phase II enzyme induction, inflammation, mitochondrial perturbation
Mitochondrial perturbation	Inner membrane damage, permeability transition (PT) pore opening, energy failure, apoptosis, apo-necrosis, cytotoxicity
Inflammation	Tissue infiltration with inflammatory cells, fibrosis, granulomas, atherogenesis, acute phase protein expression (e.g. C-reactive protein)
Uptake by reticulo-endothelial system	Asymptomatic sequestration and storage in liver, spleen, lymph nodes, possible organ enlargement and dysfunction
Protein denaturation, degradation	Loss of enzyme activity, auto-antigenicity
Nuclear uptake	DNA damage, nucleoprotein clumping, autoantigens
Uptake in neuronal tissue	Brain and peripheral nervous system injury
Perturbation of phagocytic function, 'particle overload,' mediator release	Chronic inflammation, fibrosis, granulomas, interference in clearance of infectious agents
Endothelial dysfunction, effects on blood clotting	Atherogenesis, thrombosis, stroke, myocardial infarction
Generation of neoantigens, breakdown in immune tolerance	Autoimmunity, adjuvant effects
Altered cell cycle regulation	Proliferation, cell cycle arrest, senescence
DNA damage	Mutagenesis, metaplasia, carcinogenesis

because they determine the reactivity and the energy of the surface. It is the surface of nanomaterials that will interact with the external environment. Certain features, such as transition metals on the surface, can lead to excessive generation of reactive oxygen species (ROS), e.g. free radicals, which can overwhelm the natural antioxidant defences of a given biological system and then lead to severe responses and injuries. Some metals, due to their specific electronic structure and bonding properties, are also able to catalyse unwanted reactions as detailed in figure 2.12. It is therefore essential that the core inorganic nanomaterial is either not toxic, or is covered with a protective coating that will prevent it from degrading, or toxic ions leaching out into the body. For example, quantum dots with polymer coating have been shown to have improved chemical stability and are less toxic [50]. Similarly, the surface charge (measured as zeta potential) and hydrophobicity affect nanomaterial toxicity significantly. Figure 2.13 shows results from the screening of 130 nanoparticles and evaluates the toxic effects of the particle sizes (further discussed below), their zeta potential and the surface hydrophobicity. The study showed that positively charged small nanoparticles with low hydrophobicity can cause severe cytotoxicity, due to a combination of their strong interactions with various biological components as well as their inability to 'clear-out' of the system. Table 2.2 gives an extensive list for advanced readers of nanomaterials of varying chemical compositions and the toxicity caused by them. In most cases, potential strategies to mitigate the toxicity include some form of surface passivation, by either tethering biocompatible molecules on the surface, reducing the surface charge and/or modifying the surface chemistry/structure [51].

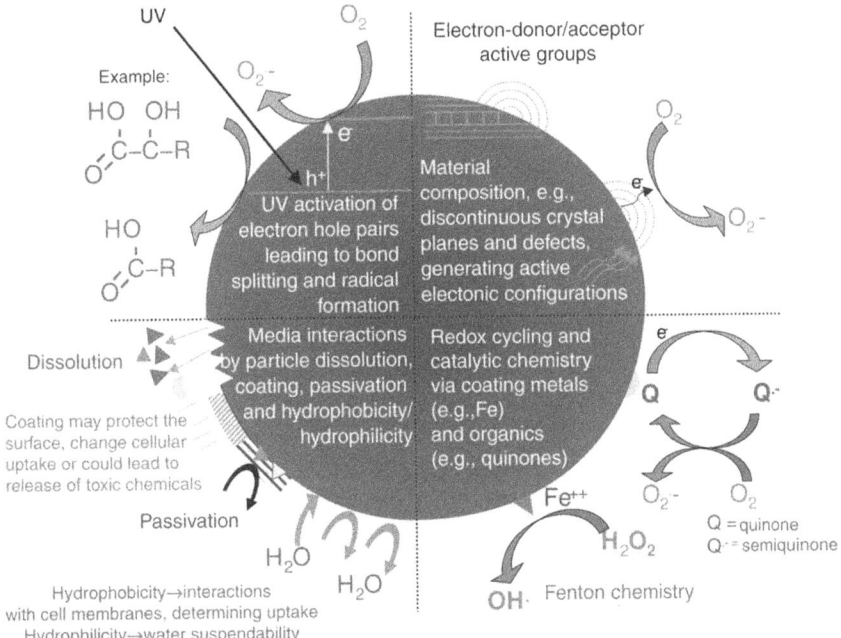

Figure 2.12. Pathways and mechanisms of nanomaterial–biological tissue interactions. Image from [46], reprinted with permission from AAAS.

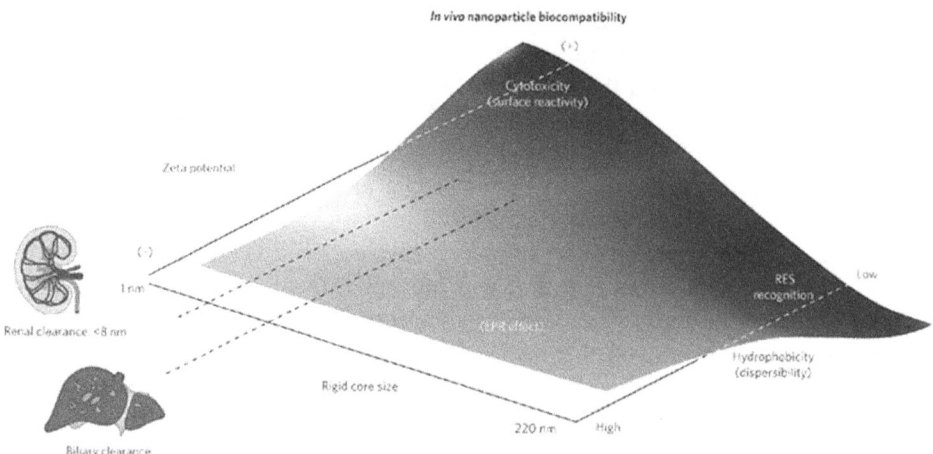

Figure 2.13. A 3D qualitative phase diagram showing biocompatibility trends as a function of the properties of nanoparticles using data from *in vivo* screening of around 130 nanoparticles. Red = likely toxic, blue = likely safe, and blue–green–yellow = intermediate levels. RES = reticuloendothelial system and EPR = enhanced permeation and retention. Image taken from [51], copyright 2009 with permission of Springer.

Physical features: Particles and materials in the nanoscale influence and interact at the cellular, subcellular and protein levels given the overlap of the length scales (figure 2.1). Examples include wrapping proteins or DNA on the nanoparticle surface, disrupting protein–protein assemblies by the insertion of nanoparticles, and disrupting lipid–bilayer structures in cell walls. Further, given the size, nanoparticles can travel through the body and penetrate various membranes, causing deposition or accumulation in certain areas, thereby triggering biological (e.g. immune) responses. A key physical feature of nanomaterials that strongly influence their toxicity is the total available surface for interactions with biological systems. As we have seen in figure 2.2, with the size decreasing, the surface area increases for a constant volume of particles. This allows significantly more opportunities for interactions with cellular and subcellular components, thus enhancing the toxic effects.

Some of the physical effects can be illustrated by discussing a classic case of silicates/silica. As discussed in the section above, silica in its amorphous phase is typically not toxic, even at nanoscales, and hence it has been 'generally regarded as safe (GRAS)' by regulating bodies in the USA and Europe for use in food and cosmetic applications. However, quartz, a crystalline form of silica, of any shape, causes severe lung inflammation, even at low exposures and at μm size [47]. The crystalline nature of the surface induces the formation of ROS and oxidative stress in biological systems. Asbestos, a naturally occurring fibrous silicate mineral, is well known for its toxicity, leading to fatality. It is the shape of the asbestos that causes its adverse effects and not the chemical composition. The nanoscale diameter of the fibres and microscale lengths mean that asbestos can find its way into lungs, but can get easily trapped. This peculiar shape makes some of the immune responses ineffective, leading to scarring (asbestosis).

Table 2.2. A list of chemically distinct nanomaterials with their cytotoxicity. Table adapted from [51].

Nanomaterial	Cytotoxicity mechanism
TiO_2	ROS production mediated by electron–hole pairs. Glutathione depletion and toxic oxidative stress as a result of photoactivity and redox properties. Nanoparticle-mediated cell membrane disruption lead to cell death and protein fibrillation
ZnO	ROS production. Dissolution and release of toxic cations. Lysosomal damage. Inflammation
Ag	Dissolution and Ag^+ release inhibits respiratory enzymes and adenosine triphosphate (ATP) production. ROS production. Disruption of membrane integrity and transport processes
Au	Disruption of protein conformation
CdSe	Dissolution and release of toxic Cd and Se ions
SiO_2	ROS production by surface defects in crystalline silica and impurities in silicates. Protein unfolding and membrane disruption
Fe_3O_4	ROS production and oxidative stress. Liberation of toxic Fe^{2+}. Disturbance of the electronic and/or ion transport activity in the cell membrane
CeO_2	Protein aggregation and fibrillation
SWCNT, MWCNT	Frustrated phagocytosis causes chronic tissue inflammation and DNA oxidative injury. Generation of ROS due to the metal impurities trapped inside carbon nanotubes (CNTs). Pro-inflammatory effects due to oxidant injury. Granulomatous inflammation due to hydrophobic CNT aggregation. Interstitial pulmonary fibrosis due to fibroblast-mediated collagen production
Fullerenes	ROS production (spontaneous or photoactivated). Hydrophobic surface increases aggregation but promotes intramembranous localization
Dendrimers	Membrane damage, thinning and leakage. Damage to the acidifying endosomal compartment by the proton sponge effect that allows entry into the cytosol
Al_2O_3	ROS production. Pro-inflammatory response
Cu/CuO	DNA damage and oxidative stress
MoO_3	Membrane disruption

Bioavailability: This includes the fate of nanomaterials in the body which depends on aspects such as the solubility/biodegradation of nanomaterials and the body's ability to clear them via various mechanisms (e.g. see figure 2.13). The exposure levels and the routes of exposure are also important factors affecting bioavailability. If certain particles, which can be toxic, are readily removed/excreted by the body then they offer little threat. Similarly, if nanoparticles are able to break down/dissolve *in vivo*, they can be safe. For example, in addition to the shape effect, the inability of asbestos to dissolve or break down poses a sustained stress/injury, while amorphous silica can dissolve in the body and can be easily removed [52]. On the other hand, dissolution of particles where the toxic core is protected by biocompatible coatings can induce toxic responses. In such cases, the anticipated fate is removal by the body before any leaching of the core happens.

2.6 Summary: key lessons from nanomaterials, nanoproperties and applications

Summary of content

- Nanomaterials are defined as having at least one dimension in the nanoscale range, be this nanosheets (one dimension), nanowires (two dimensions) or nanoparticles (three dimensions). It should be noted that nanoporous materials are also considered nanomaterials due to many shared properties.
- The breadth of types of nanomaterials, nanoproperties, and application for nanomaterials are too great to cover even superficially in one chapter. Thus this book only considers inorganic nanomaterials, and mainly focuses on nanoparticles.
- In this chapter we have focused on the optical and magnetic properties of mainly gold, silver, CdSe, and magnetic iron oxide and considered mainly nanomedical applications of these materials. Nanodevices and consumer products of a range of other nanomaterials are also considered briefly.

Key lessons

- Nanomaterials have a very high surface area-to-volume ratio. This means surface chemistry and physics dominates their properties.
- The large surface area is also useful for increasing the activity of surface reactions (i.e. heterogenous catalysis) and ideal for delivery of an active species attached to the surface (i.e. high loading of drugs, etc).
- Small size also lends itself to nanomedical applications as it warrants access to everywhere in the body and it also makes nanomaterials a comparable size to biological targets such as proteins.
- Nanomaterials lie at the boundary between atomic level quantum physics and bulk material Newtonian physics, and thus exhibit properties between the two.
- Surface plasmon resonance (SPR) is a surface effect of how an electromagnetic wave polarises and thus oscillates free electrons on metal surfaces. This absorbs and scatters light dependent on the polarizability of the electrons in that material, the size and shape of the particle and the surrounding media.
- SPR can be used to heat a particle where the energy absorbed is converted to phonons (lattice heat) and this can provide photothermal therapies when the particles are irradiated with a laser.
- SPR has proved very useful for increasing the sensitivity of nanomedical sensing and has also been used to create tools such as molecular rulers.
- Quantum confinement is a property of semiconductors which occurs when the particle becomes comparable to or smaller than its Bohr radius. At this size the electrons are confined in all directions which results in a bigger band gap (proportional to the radius of the particle) and strong fluorescence at this energy. These particles are quantum dots.

- Quantum dots have proved very useful for *in vitro* imaging. They are ideal as they have very intense fluorescence which does not fade, and the colour can be tuned with materials and particle size. Concerns over toxicity of several semiconductor materials such as CdSe quantum dots means they are slower to be used *in vivo*.
- At the nanoscale, magnetic particles are single domain and just before they become multidomain they have the highest magnetic coercivity for that material. This is because there is no energy wasted in creating the domain wall. As the size decreases they become superparamagnetic, which means the particles exhibit no coercivity.
- Iron oxide magnetic nanoparticles are ideal for nanomedical therapies. The magnetic properties can be used to magnetically target the particles and their payload to a site in the body using a magnetic field gradient. The magnetic nanoparticles can be heated by application of an alternating magnetic field. The heat is provided from the internal magnetic resistance to the flipping field for magnetic nanoparticles with coercivity, or the external friction of a superparamagnetic particle against their media as they flip. The heating can provide hyperthermic therapies. Magnetic nanoparticles can be used to enhance MRI imaging by providing a local magnetic field to increase relaxation times, and superparamagnetic iron oxide nanoparticles can be used in magnetic particle images which has increased intensity by millions of times compared to MRI. Magnetic nanoparticles can also be used *in vitro* for cell separation and sample concentration for biosensing.
- Nanotoxicity and biocompatibility must be considered when designing all types of nanomaterials, not just those for biomedical use.
- Nanotoxicity must be considered throughout the whole life cycle of the production of the material, the use of the material and the disposal of the material.
- The section on nanodevice technology demonstrates the real power sophisticated nanomaterials platforms have to revolutionise technology, from personalised point-of-care diagnostic healthcare chips, to ultra-high-density data storage and spintronic electronics of the future.
- A broader range of materials are discussed in consumer products to give a flavour of the breadth of uses for various nanomaterials.
- As one of the largest and expanding fields in nanoscience, nanomedicine was the focus of the applications section. In medicine and healthcare we are seeing incredible developments of smart and sophisticated multifunctional nanomaterials.
- Across the applications it is clear that as the design of our nanomaterials becomes more sophisticated, we are challenged to see if the synthesis of more intricate nanomaterials with more demanding specifications can keep up and deliver these materials precisely and consistently. This is ever more important for nanomaterials for *in vivo* medicines. These need to be precise and consistent, but also produced in a non-toxic way that ensures the nanomaterials are biocompatible and safe for medical use.

References

[1] Erbas-Cakmak S, Leigh D A, McTernan C T and Nussbaumer A L 2015 *Chem. Rev.* **115** 10081
[2] Eustis S and El-Sayed M A 2006 *Chem. Soc. Rev.* **35** 209
[3] Lal S, Link S and Halas N J 2007 *Nat. Photonics* **1** 641
[4] Anker J N, Paige Hall W, Lyandres O, Shah N C, Zhao J and Duyne R P V 2008 *Nat. Mater.* **7** 442
[5] Hu M, Chen J, Li Z-Y, Au L, Hartland G V, Li X, Marquez M and Xia Y 2006 *Chem. Soc. Rev.* **35** 1084
[6] Wang Y, Hu R, Lin G, Roy I and Yong K-T 2013 *ACS Appl. Mater. Interfaces* **5** 2786
[7] Blundell S 2001 *Magnetism in Condensed Matter* (Oxford: Oxford University Press)
[8] El-Sayed I H 2010 *Curr. Oncol. Rep.* **12** 121
[9] Piao J-G, Gao F, Li Y, Yu L, Liu D, Tan Z-B, Xiong Y, Yang L and You Y-Z 2018 *Nano Res.* **11** 3193
[10] Meng Q-F *et al* 2018 *Nanotechnology* **29** 134004
[11] Bobo D, Robinson K J, Islam J, Thurecht K J and Corrie S R 2016 *Pharm. Res.* **33** 2373
[12] El-Sayed I H, Huang X and El-Sayed M A 2005 *Nano Lett.* **5** 829
[13] Piludu M, Lantini M S, Cossu M, Piras M, Oppenheim F G, Helmerhorst E J, Siqueira W and Hand A R 2006 *Arch. Oral Biol.* **51** 967
[14] Scotuzzi M, Kuipers J, Wensveen D I, de Boer P, Hagen K W, Hoogenboom J P and Giepmans B N G 2017 *Sci. Rep.* **7** 45970
[15] Cang H, Sun T, Li Z-Y, Chen J, Wiley B J, Xia Y and Li X 2005 *Opt. Lett.* **30** 3048
[16] Yu E Y, Bishop M, Zheng B, Matthew Ferguson R, Khandhar A P, Kemp S J, Krishnan K M, Goodwill P W and Conolly S M 2017 *Nano Lett.* **17** 1648
[17] Thi T T H, Tran D-H N, Bach L G, Vu-Quang H, Nguyen D C, Park K D and Nguyen D H 2019 *Pharmaceutics* **11** 120
[18] Alphandéry E, Chebbi I, Guyot F and Durand-Dubief M 2013 *Int. J. Hyperth.* **29** 801
[19] Lee J, Kim H, Kim S, Lee H, Kim J, Kim N, Park H J, Choi E K, Lee J S and Kim C 2012 *J. Mater. Chem.* **22** 14061
[20] Goel S, England C G, Chen F and Cai W 2017 *Adv. Drug Deliv. Rev.* **113** 157
[21] Jun Y-W *et al* 2005 *J. Am. Chem. Soc.* **127** 5732
[22] Muthana M *et al* 2015 *Nat. Commun.* **6** 8009
[23] Gautier J, Allard-Vannier E, Burlaud-Gaillard J, Domenech J and Chourpa I 2015 *J. Biomed. Nanotechnol.* **11** 177
[24] Yu M K, Kim D, Lee I-H, So J-S, Jeong Y Y and Jon S 2011 *Small* **7** 2241
[25] Ulbrich K, Holá K, Šubr V, Bakandritsos A, Tuček J and Zbořil R 2016 *Chem. Rev.* **116** 5338
Mou X, Ali Z, Li S and He N 2015 Applications of magnetic nanoparticles in targeted drug delivery system *J. Nanosci. Nanotechnol.* **15** 54
[26] Dobson J 2006 *Nanomedicine* **1** 31
[27] Pankhurst Q A, Connolly J, Jones S K and Dobson J 2003 *J. Phys. D - Appl. Phys.* **36** R167
[28] Li Z, Kawashita M, Araki N, Mistumori M and Hiraoka M 2011 *Bioceram. Dev. Appl.* **1** 1
[29] Das P, Colombo M and Prosperi D 2019 *Colloids Surf.* B **174** 42
[30] Kallumadil M, Tada M, Nakagawa T, Abe M, Southern P and Pankhurst Q A 2009 *J. Magn. Magn. Mater.* **321** 1509

[31] Sangnier A P, Preveral S, Curcio A, Silva A K A, Lefèvre C T, Pignol D, Lalatonne Y and Wilhelm C 2018 *J. Control. Release* **279** 271
[32] Hergt R, Dutz S, Müller R and Zeisberger M 2006 *J. Phys. Condens. Matter.* **18** S2919
[33] Hergt R, Hiergeist R, Zeisberger M, Schuler D, Heyen U, Hilger I and Kaiser W A 2005 *J. Magn. Magn. Mater.* **293** 80
[34] Vines J B, Yoon J-H, Ryu N-E, Lim D-J and Park H 2019 *Front. Chem.* **7** 167
[35] Wang Q, Liu R, Yang X, Wang K, Zhu J, He L and Li Q 2016 *Sens. Actuators* B **223** 613
[36] Sönnichsen C, Reinhard B M, Liphardt J and Alivisatos A P 2005 *Nat. Biotechnol.* **23** 741
[37] Jung W, Han J, Choi J-W and Ahn C H 2015 *Microelectron. Eng.* **132** 46
[38] Dobisz E A, Bandic Z Z, Wu T and Albrecht T 2008 *Proc. IEEE* **96** 1836
[39] Yan W, Txoperena O, Llopis R, Dery H, Hueso L E and Casanova F 2016 *Nat. Commun.* **7** 13372
[40] Lafleur J P, Jönsson A, Senkbeil S and Kutter J P 2016 *Biosens. Bioelectron.* **76** 213
Samiei E, Tabrizian M and Hoorfar M 2016 *Lab. Chip* **16** 2376
Syedmoradi L, Daneshpour M, Alvandipour M, Gomez F A, Hajghassem H and Omidfar K 2017 *Biosens. Bioelectron.* **87** 373
[41] Thomson T and Terris B D 2011 *Developments in Data Storage* (New York: Wiley), p 256
[42] Chong T C and Piramanayagam S N (ed) 2011 *Developments in Data Storage: Materials Perspective* (Piscataway, NJ: Wiley-IEEE Press), p 352
[43] Premasiri K and Gao X P A 2019 *J. Phys. Condens. Matter.* **31** 193001
[44] Bhatti S, Sbiaa R, Hirohata A, Ohno H, Fukami S and Piramanayagam S N 2017 *Mater. Today* **20** 530
[45] Vance M E, Kuiken T, Vejerano E P, McGinnis S P, Hochella M F, Rejeski D and Hull M S 2015 *Belstein J. Nanotechnol.* **6** 1769
[46] Nel A, Xia T, Mädler L and Li N 2006 *Science* **311** 622
[47] The Royal Society and The Royal Academy of Engineering 2004 Nanoscience and nanotechnologies: opportunities and uncertainties, Joint report https://royalsociety.org/~/media/Royal_Society_Content/policy/publications/2004/9693.pdf
[48] Stohs S J and Bagchi D 1995 *Free Radic. Biol. Med.* **18** 321
[49] Farooq M A and Dietz K J 2015 *Front. Plant Sci.* **6** 14
[50] Hoshino A, Fujioka K, Oku T, Suga M, Sasaki Y F, Ohta T, Yasuhara M, Suzuki K and Yamamoto K 2004 *Nano Lett.* **4** 2163
[51] Nel A E, Mädler L, Velegol D, Xia T, Hoek E M V, Somasundaran P, Klaessig F, Castranova V and Thompson M 2009 *Nat. Mater.* **8** 543
[52] He Q, Shi J, Zhu M, Chen Y and Chen F 2010 *Micropor. Mesopor. Mat.* **131** 314

Chapter 3

Characterisation of nanomaterials

3.1 Introduction

There are numerous methods for characterising materials, based on their chemical composition and structure as well as their chemical and physical properties and functions. Chemical composition can be interrogated by mass spectrometry and inductively coupled plasma spectroscopy (ICP), while structure can be determined by powder x-ray diffraction (XRD), nuclear magnetic resonance (NMR), and crystallography. Chemical properties can be found by infra-red (IR) spectroscopy, UV–vis, and physical properties assessed by porosimetry and stress–strain measurements. The key to characterisation and choosing the right technique is to ensure the characterisation technique is answering the characterisation question that you are asking. Critically for nanomaterials, all the methods above give information about the whole material, and no information specifically at the nanoscale. Therefore, the birth of nanoscience and engineering has developed hand in hand with nanoscale characterisation techniques (particularly microscopy), as one cannot progress without the development of the other. Namely, until one is able to 'see' at the nanoscale we cannot begin to develop experiments to utilise and understand nanomaterials. Figure 3.1 shows schematically some commonly used characterisation techniques and the features of materials they can probe. Typical examples of materials/systems in each length scale are also shown.

Thus, the development of electron microscopy has been essential to initiating nanoscience. In this section we will briefly describe characterisation specifically for nanoscale features, such as electron and force microscopy, nanoscale specific spectroscopy and scattering techniques. It should however be noted that these techniques can and should be complemented by other characterisation techniques (such as those listed at the beginning of this section) to give a full understanding of the material and its properties.

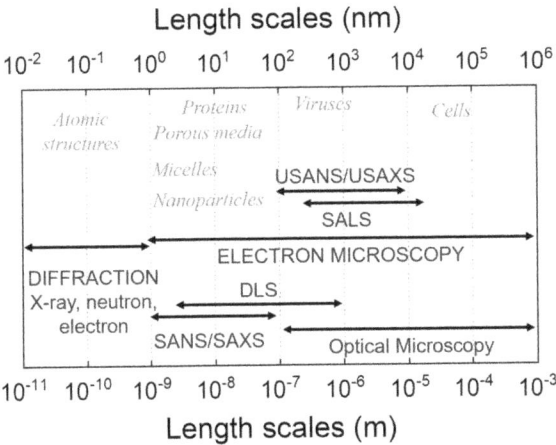

Figure 3.1. Schematic showing the different characterisation techniques and the length scales/materials they probe.

3.2 Microscopy

3.2.1 Optical microscopy

Most people equate microscopy with optical microscopy, where a sample is viewed on an illuminated stage (with either natural or artificial light) using a series of lenses to bend the light and magnify the sample's features. While light microscopy is an excellent characterisation technique, it has a physical limit to its resolution, which is the wavelength of light. The diffraction limit ($d = \lambda/2$) means that when measured at wide angles, light cannot be used to observe something smaller than half its own wavelength (≈250 nm for green light) [1]. In this respect optical microscopy cannot be considered a characterisation method for nanoscale materials. However, over the last 30 years there has been a concerted effort to surmount this physical limit, and to develop super-resolution light microscopy. There have been several approaches, and some of these have been combined. One approach exploits the evanescent wave produced by confining excitation laser light through a hole smaller than its wavelength. This can be used to visualise a sample with the resolution dependent on the size of the aperture (hole), not the wavelength, if held very close to (a few nm away) from the sample surface. This technique, near-field scanning optical microscopy (NSOM or SNOM), can give resolutions down to 20 nm [2]. However, the close distance to the sample surface and relatively large probe means there are limitations to the sort of samples that can be imaged, furthermore there is a very shallow depth of field, and the method can have very long scan times.

Other methods of super resolution light microscopy use fluorescence microscopy. By performing fluorescence microscopy, but only on optically distant fluorescent probes (so they can be resolved), a resolution of 200 nm can be achieved in spectral precision distance microscopy (SPDM) [3]. Similarly, the signal generated from a dye bound to DNA can be used in binding activated localization microscopy (BALM) to image DNA in the nucleus, while a fluorescent probe in cold conditions to slow down the fluorophores chemistry can be used to obtain molecular level resolution (if compared/combined with structural information from election microscopy or crystal structure data) in cryogenic optical localisation in 3D (COLD) [4]. Further to this, several techniques have been developed that visualise fluorescent probes at different times using photoswitchable fluorophores. Thus, only distant fluorophores are resolved at each time point, allowing a high-resolution image to be built up over time. Techniques include stochastic optical reconstruction microscopy (STORM), and (fluorescence) photo active localization microscopy (PALM and FPALM), which along with sophisticated computational deconvolution now have resolution limits down to around 30 nm.

Advantages of optical microscopy are that it can use liquid ambient environments, can be dynamic and has a huge size range. However, samples usually need to be tagged with specific probes for this method and the resolution, while improving, is still not adequate for imaging many native nanoscale features.

3.2.2 Electron microscopy

Another way to overcome the diffraction limitation of light is to not use light as the vehicle for imaging. Election microscopy uses accelerated electrons to visualise the sample, similar to the way photons (light) are used in optical microscopy. Accelerated electrons have a wavelength much smaller than that of visible light (2.5 pm (at 200 keV)), resulting in electron imaging resolution being practically unrestricted by the wavelength. Additionally, electrons interact more strongly with matter than photons, so a stronger signal can be obtained from samples with reduced dimensions (figure 3.2). The source of electrons for imaging is from an elemental solid target such as a a tungsten filament, which releases electrons on heating (thermionic emission) or by quantum tunnelling when exposed to a strong electric field (field emission). However, it is becoming more common to use a Schottky-style field emission gun as the source of electrons in modern transmission electron microscopes. The electron beam is then focused down the column onto the sample using a series of lenses. These are analogues to optical glass lenses, but are in this case electromagnetic lenses, using the electromagnetic field to direct and focus the charged electrons (see figure 3.3 in section 3.3.2.4).

Electron microscopy has the added advantage that the electrons can be used to provide structural and elemental information about the sample *in situ* at the same resolution of the electron beam. The electron beam can be diffracted by the sample to give a crystal diffraction pattern that can be assigned to a crystal structure. Furthermore, the electron beam can ionise the sample by 'knocking out' inner shell electrons of atoms in the sample. Electrons in higher energy levels then drop into the vacant lower energy states, potentially emitting x-rays as part of the de-excitation process, critically at the energy (wavelength) of that atomic energy level transition which is element specific. Energy dispersive x-ray analysis (EDAX) measures the energy of x-rays emitted from a sample impacted by an electron beam, to determine the atomic species that are present. As a result, the x-rays that are produced have characteristic energies and wavelengths depending on the atomic species that are

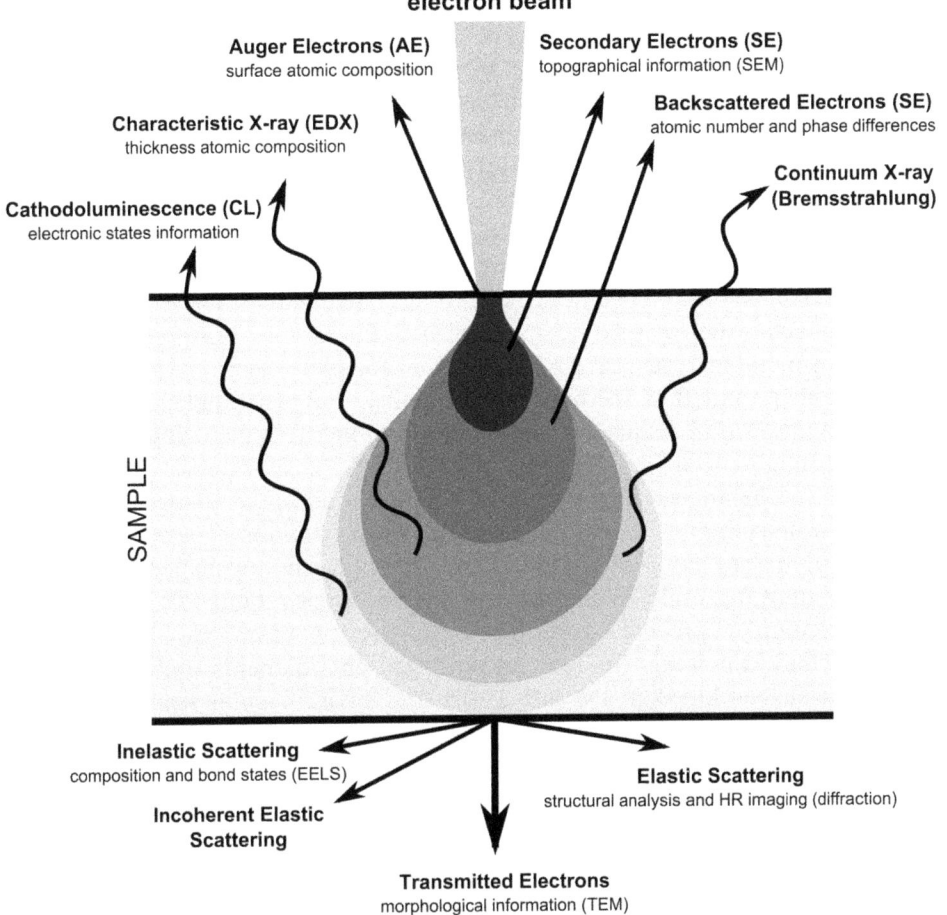

Figure 3.2. Schematic of high energy electron interactions with a sample during electron microscopy, showing the depth from which such waves can penetrate. Wikimedia Commons images © Claudionico under Creative Commons Licence.

present (figure 3.2). Electron microscopy is thus a very powerful tool with the ability to give topological, structural and elemental information at close to atomic level resolution. However, one disadvantage is that all optical paths in an electron microscope have to be held under high vacuum.

3.2.3 Scanning electron microscopy

Scanning electron microscopy (SEM) has become one of the most important and widely used techniques for the characterisation of surfaces containing nanoscale features or nanomaterials on a surface. SEM gives an image of the topology of the surface (figure 3.2) (i.e. it does not penetrate the sample) as the beam is scanned across it in a raster scan. Primary electrons are accelerated (approx. 0.1–40 keV) and focused onto a sample to interact with a volume that is dependent on the energy of the primary beam, and the atomic number of the sample. After electron–surface interactions, secondary electrons escape from the sample with a kinetic energy <50 eV. Secondary electrons are most likely ionised electrons from atoms close to the sample surface, or primary electrons that have lost almost all their energy through scattering. Backscattered electrons are also produced as a result of primary electrons undergoing large deflections, leaving the surface with only small changes to their initial kinetic energy. Furthermore, atoms within the sample can undergo inner shell ionisation which can be utilised to collect (by EDAX) chemical information about the sample, as discussed above (figure 3.2). Electric charge will thus build up on non-conductive samples, so these need to be coated with a thin layer of an inert conducting metal, such as gold or silver, to earth the sample. Furthermore, coating with a heavy metal will increase the signal of the secondary electrons (as an increased atomic number) and thus increase image quality, but will hinder EDAX, so if elemental analysis is necessary no coating or a non-interfering coating (e.g. carbon) is required.

Samples can be imaged in a variety of different modes, but topographic images obtained by detecting secondary electrons are most common. These are normally detected with a scintillator–photomultiplier system placed at a shallow angle to the side of the imaged sample. While electron microscopy is required to be performed under a vacuum to prevent scattering of the electron beam, an SEM has been developed that can image in low pressure gas environments and humid/liquid environments. Developed in the late 1980s, the environmental scanning electron microscope (ESEM) can image wet samples in atmospheres by separating the beam column (kept under vacuum) and the sample chamber (in the environmental condition), and using a secondary electron detector modified to collect secondary electrons in humid environments. This is a very important advancement for SEM, particularly for the analysis of biological samples. First, a conductive coating is not needed, allowing better image detail and easier chemical analysis, and second, samples can be observed in a range of environments, such as a biological sample in its native environment.

3.2.4 Transmission electron microscopy

Transmission electron microscopy (TEM) is similar in principle to SEM, and has also become one of the key techniques available for imaging very fine features and ultrathin structures on the nanoscale. TEM operates at much higher accelerating voltages than SEM (typically 100–400 keV, but can be up to 1 MeV), ultimately leading to greater resolution. It is now routine to resolve crystal lattices and even atomic level features. As with SEM, a TEM focuses electrons through the application of electric or magnetic fields onto a sample in high vacuum conditions (figure 3.3). However, this sample must be thin (no more than a few nanometres for hard materials, <1 μM for softer samples), as the technique relies on the electrons being transmitted *through* the sample (figure 3.2) and are detected underneath the sample with a phosphor screen or a charged coupled device to form a two-dimensional projection of the sample. A TEM can image in a number of different

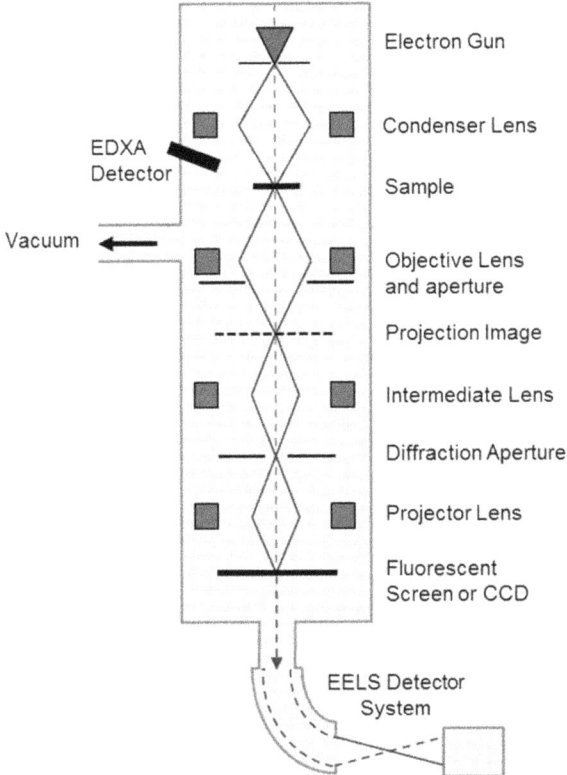

Figure 3.3. Schematic showing the typical layout of a TEM. Electrons are focused onto a sample through the use of electromagnetic lens systems (red boxes) under high vacuum conditions. After passing through the sample, the electron beam (blue line) is then focused to form an image. The typical locations for EDXA and EELS detection systems are also shown. Adapted from [5], copyright of Oxford University Press 2000.

modes; a bright field image can be generated as electrons undergo a number of different scattering processes when interacting with the sample. Regions that are more dense or thicker will lead to greater scattering, which will not be detected and lead to these regions appearing darker on the final image. Crystalline samples also diffract the electron beam, which can be used to form a dark field image if the transmitted beam is excluded. Larger soft samples can be imaged by thin sectioning, which is where the sample can be set in a resin then sliced into thin sections (50–200 nm thick depending on requirements). Similar to SEM, samples with low atomic number (particularly biological samples) can be stained with a heavy metal solution such as uranyl acetate to increase the contrast. Also similar to SEM, there have been recent advances in the development of methods to image samples within a liquid/gas environment. Fluid cell TEM (or environmental TEM/*in situ* TEM) again utilises a separate sample chamber so the electron beam column can be retained under high vacuum, while the sample chamber itself comprises a very thin chip with a fluid chamber, with an electron invisible nitride window above and below the sample for the electrons to be transmitted through. Again, using such systems, living biological cells and chemical reactions such as nanoscale crystallisation events can be observed in nanoscale detail in real time.

3.2.5 Atomic force microscopy

Like electron microscopy, scanning probe techniques have become one of the most well used methods for characterising surfaces and nanomaterials with nanoscale resolution. Atomic force microscopy (AFM) is a topological method that is able to map a surface or nanoscale feature on a surface by reading the force of attraction/ repulsion between a probe and the surface (figure 3.4). This simple method means that AFM has the ability to image materials without the need for high vacuum conditions, and has become one of the most versatile characterisation tools in nanoscience. AFM images the topography of a sample by scanning an extremely sharp tip (the sharper the tip, the better the resolution) mounted on the free end of a cantilever over the surface in sequential line scans. Several forces act between the tip and the surface, causing the cantilever to deflect. The most common are attractive van der Waals forces, when the tip is between 10–100 Å from the surface, and repulsive electrostatic forces, when the tip is <2 Å from the surface. A topographical image of the scanned sample is built from measurements of the cantilever positions, usually with a laser beam that is reflected off the back of the cantilever and detected with a position-sensitive photodetector. AFM can be operated in a variety of different modes, including constant-height, where the scanner height is fixed and an image is built from measurements of the cantilever deflection, and constant-force, where measurements of the cantilever deflection are used in a feedback loop to maintain it at a constant force by moving the scanner over the surface topology. AFM can also be operated in non-contact modes, such as the tapping mode, where a

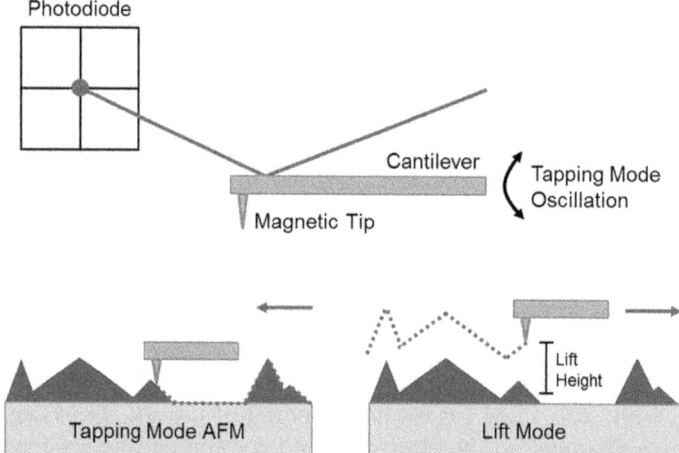

Figure 3.4. Schematic of AFM and MFM. The defection of an AFM cantilever is detected by the reflection of a laser off the cantilever and onto a photodiode. The surface is first imaged with tapping mode AFM, before the cantilever is raised a certain lift height and the surface is retraced to record magnetic information. Reproduced with permission from [6, 7].

stiff cantilever is vibrated at close to resonant frequency so that the tip just touches the sample surface at the bottom of its movement. When the tip approaches the surface it causes a shift in the frequency of vibration, but with the aid of a feedback system the AFM vibrates the cantilever at a constant frequency by moving the scanner up and down. This provides a method of recording surface topography, but due to the fact the tip is not dragged laterally across the surface, it causes less damage than contact mode AFM.

Furthermore, the tip can be functionalised to obtain more information. For example, it can be magnetised, to not only assess the topology of a sample but also to characterise the magnetism at high locality resolution. Magnetic force microscopy (MFM) first uses tapping mode to obtain a topological image of the sample, before being retraced in lift mode at a constant lift height of 50–200 nm above the surface. The magnetic interactions between the magnetised tip and the sample cause vertical frequency to be felt by the cantilever's oscillation, which can be mapped as repulsive and attractive magnetic interactions over the sample surface. Additionally, the tip can be chemically functionalised with a molecule that can interact specifically with a certain chemistry on the surface, again giving highly localised specific chemical information. The tip can be used to measure forces: if the tip is functionalised with a molecule that binds to a different molecule on a surface, the force required to pull them apart can be measured (single molecule force microscopy); rheology and elasticity can be probed observing deformation under different applied forces, and so on.

3.3 Spectroscopy applied to nanomaterials

Spectroscopic and spectrometric techniques are generally applied to bulk materials. However, with modifications, they can be used specifically to probe surfaces within μm to nm depths. As nanomaterials have a much greater proportion of the material on their surface, these surface-specific spectroscopic methods become powerful in characterising nanomaterials. In this section, selected methods are considered. The aim is to provide a brief overview of the technique, an explanation of how it is valuable in characterising nanomaterials, and a suitable example to illustrate the discussion.

3.3.1 Mass spectrometry

A specific type of mass spectrometry—time-of-flight secondary ion mass spectrometry (ToF-SIMS)—has been developed to investigate the chemical composition and functionality of topmost layers (~1 nm). In the case of nanomaterials, ToF-SIMS provides key information on functional performance of a given nanomaterial as well as its potential toxicity. Examples include the assessment of surface chemistry to predict cell-binding and proliferation for a bioglass implant, stability of zinc oxide nanoparticles in a cosmetic formulation, and selective capture of carbon dioxide on amine functionalised silica.

In mass spectrometry, a sample surface is subjected to an ion beam, which removes molecules or more typically charged fragments of molecules (called

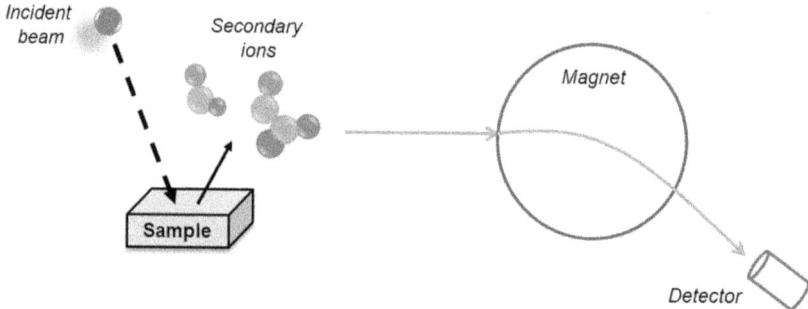

Figure 3.5. Schematic showing an incident ion beam generating secondary ions from a sample, which travel through a magnet before being collected by the recorder. Image adapted from [5], copyright of Oxford University Press 2000.

secondary ions). These secondary ions are made to travel through a varying magnetic field before being collected by a detector (see figure 3.5). Typically the time required for the flight of secondary ions to the detector is also measured, and the technique is therefore called time-of-flight secondary ion mass spectrometry, or ToF-SIMS. The output of such measurements provides a relative abundance of ions collected using the following relationship:

$$\frac{m}{z} = \frac{B^2 R^2}{2V} \tag{3.1}$$

where m and z are the mass and charge of the secondary ion, V is the acceleration potential of ions, B is the magnetic flux density of the magnet and R is the radius of the arc in which the secondary ions move. The varying magnetic field enables the separation of the secondary ions based on their charge and size (m/z). This information, when analysed for the origin of the secondary ions, can provide information on the surface chemistry of nanomaterials. One point to note is that the secondary ions can undergo breakage during their flight and result in fragments that are detected. For a multifunctional sample, the analysis can become complex due to overlapping signals from fragments of various secondary ions. This technique is less widely used for characterising nanomaterials on their own but is powerful in probing the interface between nanomaterials and a foreign molecule such as a reactant in catalysis, a pollutant or a protein.

Let us illustrate this with an example of a biomaterial [8], where researchers designed experiments to correlate the chemical functionality of biomaterials surfaces with cell proliferation. They created an array of materials by systematically varying the surface chemistry and then measured the molecular details using ToF-SIMS. The materials were then subjected to cell cultures. They observed that even for the same chemical precursors used for surface functionalisation, the processing affected the final surface functionalities. Further, they correlated the chemical information of the surface measured by ToF-SIMS with the proliferation of cells on those surfaces. Most importantly, it was found that the ions with tertiary-amine moiety (e.g.

$C_2H_6N^+$) were inhibiting cell growth, while oxygen-containing ions (e.g. CHO_2^-) amongst other ions were able to promote cell growth.

3.3.2 Infra-red spectroscopy

IR spectroscopy is well-known for qualitative and quantitative investigations of chemical bonds. The reason for this is that chemical bonds exhibit characteristic vibrations due to specific types of deformations. The frequencies of these vibrations match with the IR frequencies and the absorption of incident IR by a given sample occurs only in specific frequencies. The frequency of IR depends on the geometry of the molecule, the mass of the atoms involved the chemical bond (and hence the bond strength). The vibrations occur in stretching and bending modes. Examples of some of these vibrations are shown in figure 3.6.

IR spectroscopy has therefore been used for qualitative analysis such as identification of molecules or chemical functionalities present in a given sample, including a nanomaterial. This can provide information on the chemical composition of a nanomaterial and detect any impurities present. Further, as the absorbance of IR energy is proportional to the molar concentration of the bonds present, a quantitative analysis is also possible, where changes or differences in bonds of interest can be quantified. These methods are again powerful in studying interactions of nanomaterials with foreign molecules including solvents.

In order to specifically probe the surface of nanomaterials, two modifications of IR spectroscopy have been developed: attenuated total reflection Fourier transform IR spectroscopy (ATR-FTIR), and diffuse reflectance IR Fourier transform spectroscopy (DRIFTS). Both techniques depend on the IR beam penetrating a given sample (to ~1 μm), thus causing interaction with the chemical bonds, predominantly on the surface. In DRIFTS, the IR beam is scattered upon reaching the sample surface and results in diffused reflection. In ATR-FTIR, the instrumentation is slightly different. The incident IR beam is passed through a crystal of high refractive index. This results in total internal reflection and creates an evanescent wave that has the ability to extend beyond this crystal. The sample to be measured is placed on this crystal, and the extending evanescent wave is absorbed at certain frequencies by the sample. This causes a reduction (attenuation) of the wave. The

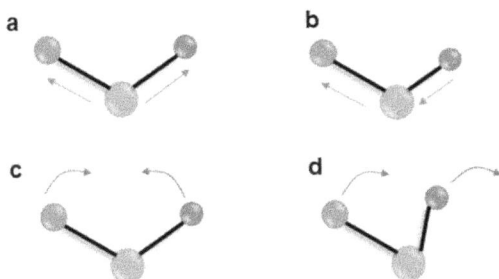

Figure 3.6. A schematic representation of stretching and bending modes in a chemical bond. (a) Symmetric stretching, (b) asymmetric stretching, (c) scissoring and (d) rocking. Reproduced from [5], copyright of Oxford University Press 2000

detector measures this attenuation of the total reflectance of the incident IR beam, which provides information on chemical bonds present in the sample. Some of the key advantages of both these techniques include the ability to probe surface and interface effects in nanomaterials and thin films. Furthermore, both require little or no sample preparation when compared to *routine* IR spectroscopy; samples can be simply placed on the holder and analysed, making this analysis rapid. Finally, ATR-FTIR can also sample liquids, which enables *in situ* monitoring of dynamic processes as well as measuring samples in their *real* environments.

Let us consider these techniques with examples, starting with ATR-FTIR. One of the key features controlling the effectiveness of a biomaterial implant is its biocompatibility. In particular, whether the host *body* accepts the implant depends on the interactions of cells with the surface of the biomaterial. However, before cells can interact, it is the proteins from the surrounding environment that adsorb on the biomaterial—depending on these proteins and how happy they are on the surface, the cells may or may not attach, which in turn decide the fate of the implant. ATR-FTIR can be used to study the protein adsorption on nanomaterials which can lead to understanding of the performance of biomaterials when implanted. One study used ATR-FTIR to investigate the interactions of bovine serum albumin (BSA) and bovine fibrinogen (Fg) on silica nanoparticles with sizes of 15–165 nm [9]. In particular, the IR spectra arising from the amide bond vibrations ($-\overset{O}{\underset{}{C}}-\underset{H}{N}-$) of the proteins can be used to infer the conformation and stability of proteins when adsorbed on nanoparticles. One of the amide bands, centred at 1700–1600 cm^{-1} as shown in figure 3.7, can be used to measure changes in the secondary structure of the protein. The components of this amide-1 band can be deconvoluted to determine the composition of α-helices, β-sheet and disordered structures within the protein structure, for example. This analysis revealed that BSA, which is a globule-like protein, was stabilised on smaller particles, while it was highly denatured when adsorbed on larger particles. On the other hand, Fg, which is rod-like in shape, retained its structure on larger particles but was denatured when adsorbed on small particles (i.e. with higher radius of curvature). This study shows how ATR-FTIR can be used to probe the interactions between nanomaterials and proteins for developing biomaterials. In particular, this work demonstrated the nanoscale effects that control protein (and in turn cell) adsorption on surfaces.

DRIFTS has found interesting applications in studying surface chemistry of a range of materials, particularly nanomaterials. Consider nanoporous carbons as an example, which find applications in catalysis and adsorption. Success depends on the porosity as well as the surface chemistry of the pores. DRIFTS was applied to probe the surface chemistry and its dynamics under heating of this technologically important porous material [10]. When used in combination with other techniques, DRIFTS was able to probe key chemical functionalities on the surface of carbons, such as carboxylic, carbonyl and phenolic.

Figure 3.8 clearly shows three peaks centred around 1750 cm^{-1}, 1600 cm^{-1} and 1250 cm^{-1}. These peaks correspond to C = O stretching (found in carboxylic anhydrides), C = O vibrations of quinone groups and C–O stretching associated

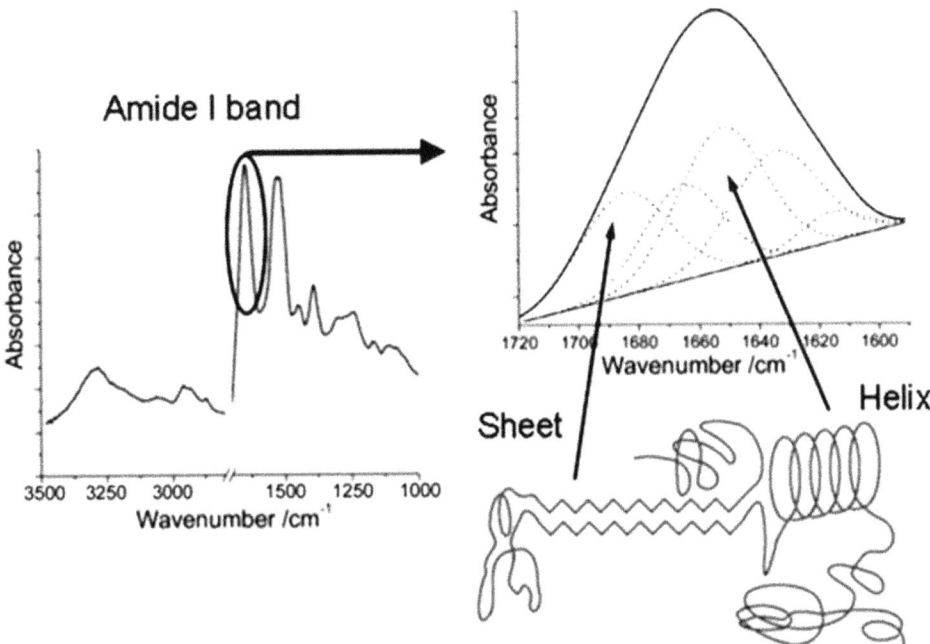

Figure 3.7. ATR-FTIR spectrum obtained for a protein (left) showing the amide-1 band of interest. The images on the right show how this band can be further analysed for underlying constituent peaks (shown with dotted lines) in order to infer secondary structures of a given protein (e.g. beta sheets and helix). Image taken with permission from [9], copyright 2006 American Chemical Society.

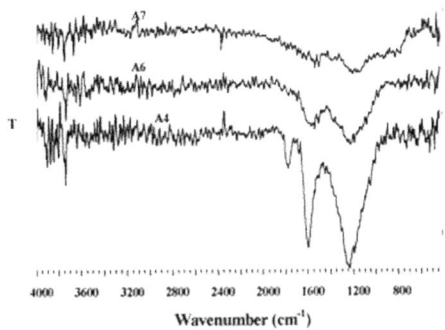

Figure 3.8. DRIFTS spectra obtained for porous carbons (sample A4 is without any treatment, A6 and A7 are heat treated samples). Image taken from [10], copyright 1999 with permission from Elsevier.

with various functionalities, respectively. It can be seen that upon heat treatment, the $C = O$ peaks progressively decrease, finally disappearing completely (sample A7). The C–O peaks, although reducing upon heat treatment, remain to some extent. These findings are not only useful in deciding on the correct samples and heat treatments needed for a given application, but they also help understand the molecular interactions controlling catalysis and adsorption.

3.3.3 X-ray photoelectron spectroscopy

Incident x-rays on a given material can cause multiple interactions and outcomes. One interaction is that the energy from the x-rays causes electrons from the top layer (a few nm deep) to escape; these escaped photoelectrons are detected and their kinetic energy is measured. The binding energy for these photoelectrons is the difference between the incident energy from the x-rays and the measured kinetic energy. The binding energy depends on the electronic configuration and the element itself, thus binding energy can be used to identify the chemical composition of the surface of a given sample. X-ray photoelectron spectroscopy (XPS) provides quantitative data and hence the chemical composition of surfaces can be fully quantified using XPS. Further, as the binding energy changes upon chemical bonding with another element, subtle shifts in binding energy can be used to study the oxidation states of elements as well and their bonding/co-ordination with other elements.

XPS can be used to understand surface chemistry in a range of applications, such as studying coatings. In the development of inorganic coatings [11], the surface was needed to be pre-coated with polyamines. A method was developed for coating surfaces with polyamines and figure 3.9(a) shows the XPS data before and after polyamine pre-treatment. The appearance of the peaks associated with binding

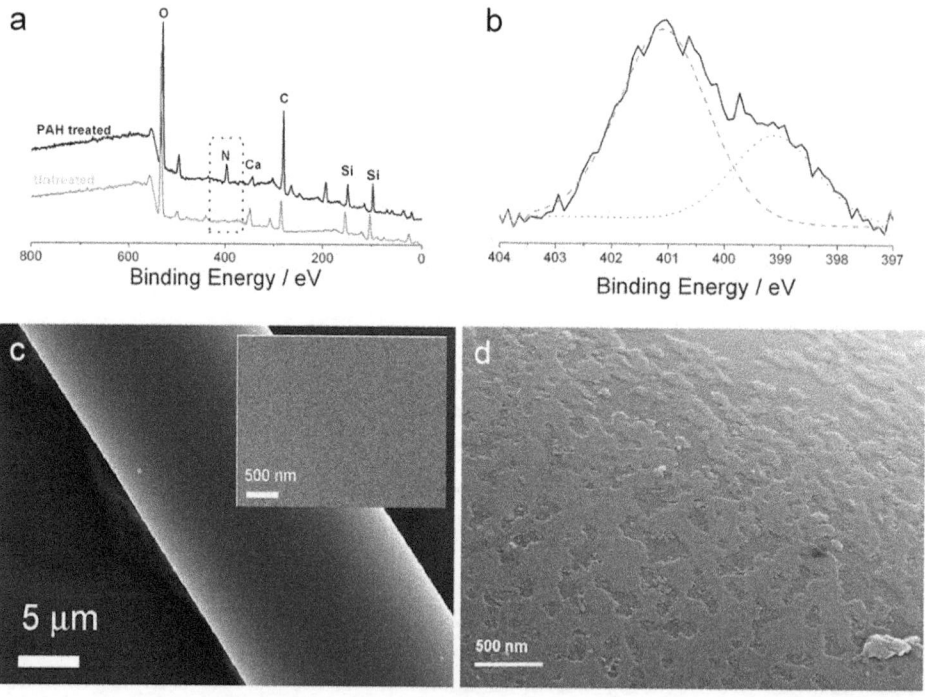

Figure 3.9. XPS (a, b) and SEM (c, d) analysis of glass-fibre surface before and after polyamine coatings. The polyamine used in this case was polyallylamine hydrochloride. Image taken with permission from [11], copyright 2007 American Chemical Society.

energy of electrons from nitrogen provided the evidence for the formation of amine coatings only in the treated sample. This was also confirmed by the SEM results, which showed smooth and coated surface before and after the polyamine treatment, respectively. The significant increase in carbon peaks also confirmed the presence of polyamine coatings. Further, as the peaks from the surface (Si, Ca and O) were also detectable, it was clear that the polyamine coating was very thin (within a few nm). Taking a closer look at the nitrogen peak (figure 3.9(b)) and performing peak deconvolution indicated that there were two sub-peaks centred at ~399 and ~401 eV. These peaks represent the amines in uncharged and charged/protonated ($\equiv N^+$) states, respectively. The quantification of the area under these peaks suggested that there were two charged sites per an uncharged site. Incidentally the charged sites were desirable for the given application. This example illustrated how XPS is a valuable tool in studying the chemical compositions of nanomaterials, all the way down to oxidation states of constituent elements. This feature makes XPS powerful in, for example, monitoring catalytic reactions where oxidation states of metal surfaces control the reaction pathways and outcomes.

3.4 Diffraction and scattering techniques

A selection of characterisation techniques that utilise how a probing electromagnetic wave (be it light or x-ray) is deflected/reflected/scattered by a sample are discussed in this section. While the light and x-ray scattering techniques give detailed information about size and shape at the nanoscale, XRD is a bulk method that gives details about the crystalline structure of any material. However, it can have specific implications for particle sizing at the nanoscale.

3.4.1 X-ray diffraction (XRD)

A crystalline material by definition is composed of a repeating ordered structure of atoms, the smallest repeat of which is called the unit cell. This regular 3D pattern of atoms can be probed, measured, and reconstructed using the constructive interference of x-rays with this crystal lattice. The Bragg father/son scientist duo won the 1915 Nobel prize for their work understanding how x-rays interact with crystalline materials and how they developed this to XRD as an analytical method.

XRD works by an incident x-ray beam striking a crystalline solid surface at an angle (θ), resulting in the production of multiple secondary reflected x-rays. In most directions these waves will cancel out due to destructive interference. However, they will add via constructive interference in certain directions based on where the atoms lay in the material and the in phase reflections these cause resulting in a diffraction pattern. This constructive interference is given by Bragg's law:

$$2d \sin \theta = n\lambda, \tag{3.2}$$

where d is the crystal lattice spacing, λ is the wavelength of the incident x-ray and n is the order (as shown in figure 3.10).

On a simple level, the x-ray hits the sample at varying angles, which comes to no consequence, but at the angle at which it hits each of the atoms that are regularly

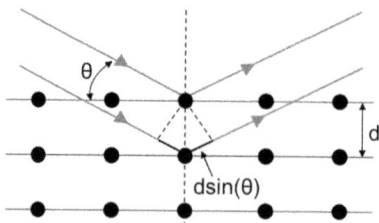

Figure 3.10. An x-ray beam incident on a symmetrical plane of atoms results in constructive interference when their path-length difference ($2d \sin \theta$) is equal to an integer number of the incident wavelength ($n\lambda$).

spaced (d-spacing) and is diffracted with their wave in phase with the other diffracted x-rays, an intense diffracted x-ray signal will be recorded. Thus an x-ray diffractometer records the intensity of the secondary reflected waves over a range of angles, either by rotating the specimen and detector while keeping the incident beam fixed, or by keeping the specimen fixed while rotating the detector and the incident beam [1]. Crystallographic planes produce characteristic peaks in the recorded spectrum at precise angles, which can be converted back to d-spacing with the use of equation (3.7); as such crystal structure can be determined from the range of d-spacings calculated, or if the material is already known, the full spectrum can be compared to a database to identify the sample. XRD can be performed on a single crystal, or powder XRD can be performed on fine powder crystalline samples. In the latter, the powder is laid on a sample plate, with an increased sample size increasing the intensity of the peaks. With XRD, the fine powder is randomly orientated giving a signature XRD pattern for any one material. As the material is orientated more in any one direction, the peak intensity of the dominant reflection will become more intense.

XRD is also a useful sizing method for crystalline nanomaterials. The average size of particles can be obtained from the peak broadening. The larger the single crystal, the more intense and narrow the peak. As the size decreases and the number of particles increases, the peak height reduces and the width broadens.

$$D = \frac{K\lambda}{\beta \cos \theta}. \tag{3.3}$$

The Scherrer equation (3.3) can be used to calculate the minimum crystallite size, where D is the size of the smallest crystallites, K is a shape factor which is normally between 1 and 0.9 for spherical particles, but does vary with different shapes, λ is the wavelength of the x-ray beam, β is the full width at half maximum of a selected peak and θ is the angle of this peak. With good quality nanoparticles this can normally be equated to the nanoparticle size. However, peak broadening can occur due to the instrument, and imperfections in the nanocrystals can give a smaller particle size, so this technique is best used in conjunction with other techniques.

3.4.2 Dynamic light scattering

Dynamic light scattering is the measurement of fluctuations observed in light scattered from a particulate sample. These fluctuations can then be related to diffusion within the sample, and then this information can be used to calculate particle size. Due to Brownian motion occurring within the sample, there is constructive and destructive interference which has an effect on the light intensity and scattering over time, this is called the translational diffusion coefficient (D) (equation (3.4)). This takes into account solution viscosity (η) Boltzmann's constant at absolute temperature (kT), and the hydrodynamic diameter (D_H) of the particle; this assumes that all particles are solid spheres.

$$D = \frac{kT}{3\pi\eta D_H}. \tag{3.4}$$

This is due to the fact that smaller particles diffuse faster through the solution, causing more fluctuations than bigger, slower diffusing particles. The technique works by way of a polarised beam, which is then scattered by density or concentration fluctuations in the sample. A detector picks up speckle caused by scattered light interfering with the sample. The time dependence of this scattering at a specified angle is used to form an autocorrelation function. For monodisperse samples this comes from the exponential decay arising from the diffusion within the sample, which is then related to the hydrodynamic diameter of the particle. If a sample is polydisperse the correlation function is based upon a distribution of decay rates. For vesicle samples a concentration of 400 µl ml^{-1} was used, with a scattering angle of 173°. Samples were generally scanned three times, with each scan having 10–14 runs. Data was analysed using Malvern Zetasizer software. It is also possible to estimate the polydispersity index (PdI) (equation (3.5)) of the sample, which relates to the standard deviation (σ) of a Gaussian distribution applied to the data (Z_D^2):

$$PdI = \frac{\sigma^2}{Z_D^2}. \tag{3.5}$$

3.4.3 Small angle scattering

Small angle scattering (SAS) is an advanced technique for studying nanoscale features of a given sample at multiple levels (e.g. primary particles, secondary particles, aggregates, etc). Depending on the energy source used (light, neutrons or x-rays), specific length scales can be mapped. As small angle x-ray scattering (SAXS) can typically probe sample features in the range of 1–100 nm, this section will focus on SAXS, although the fundamental principles apply to other SAS techniques. The scattering of incident x-rays occurs due to the difference in the electron density of a given sample. The advantages of using SAXS for characterising nanomaterials include:

- It provides a global average of properties (e.g. particle size), as opposed to local information typically obtained by microscopic techniques.
- It can probe features that are buried inside, which typically cannot be 'viewed' by microscopy and surface techniques.
- It can quantify a range of properties for nanomaterials, such as sizes of features at multiple length scales, shapes, relative concentration, polydispersity, morphology and structure.

Having said that, the analysis of SAXS data is extremely complicated, dependent on developing relevant models and requiring complementary characterisation. Therefore, this section is limited to an overview of what SAXS can offer in the context of nanomaterials, avoiding the details of the underpinning physics and the model-based analysis of the data. The length scales (or the scattering vector q) probed directly depend on the wavelength of the source and the measurement angles, as given by the following equation:

$$q = 2\pi/d = 4\pi \sin \theta/\lambda \quad (3.6)$$

where d is the feature size or characteristic length, θ is the measurement angle ($\ll 1°$ in SAS) and λ is the wavelength of the incident x-rays. This means that x-rays generated from the Cu Kα line, which has a wavelength of 0.154 nm, have the limit of detection of ~45nm if measured at an angle of 0.2°. In order to measure wider length scales, the source can be changed to lower wavelengths (or higher energy, e.g. a synchrotron radiation source) and/or the detection angles can be reduced (which requires advanced instrumentation). A typical set of data obtained from SAXS measurements shows the intensity of scattering measured at different angles (or q), as shown in figure 3.11.

Using a pre-defined model, the data is fitted to obtain parameters in the following relationship:

$$I(q) \propto q^{-\alpha} \quad (3.7)$$

where α is the fractal dimension and provides information on the shape/morphology of the sample such that the values of α of 3, 2 and 1 represent spheres, disks and rods respectively. Note that high q means smaller sizes, since q is inversely proportional to characteristic length. Figure 3.11 shows that SAXS is able to probe a wide range of length scales. Further analysis of the SAXS data shown in figure 3.11 provided sizes of primary and secondary particles, which were found to be in the range of 5–10 nm and 100s nm respectively. The analysis also identified fractal dimensions and connectivity, which provide an understanding of their formation and porous properties.

3.5 Porosimetry

The porosity of nanomaterials is of vital importance in understanding their properties as well as performance. This becomes particularly critical in applications such as catalysis and pollution control, where the effectiveness of these materials is

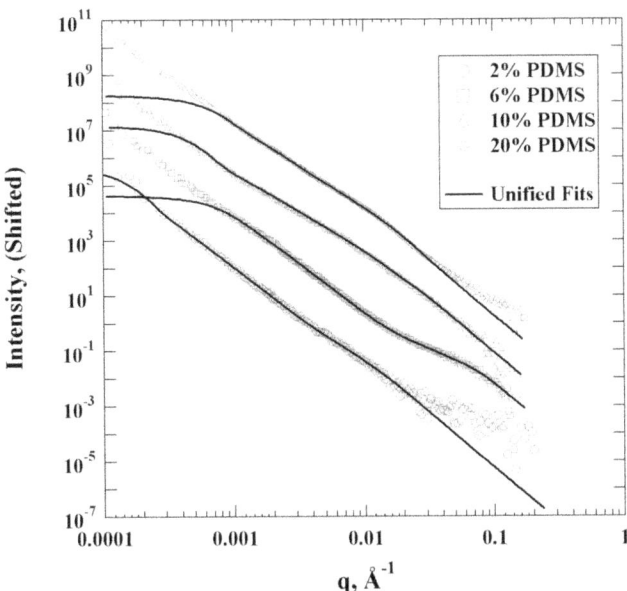

Figure 3.11. Ultra small angle x-ray scattering data obtained for a polymer filled with nanoparticles at varying amounts. Image taken from [12], copyright 2006 with permission from Elsevier.

controlled by the surface area, pore volume, pore sizes and size distribution, pore shapes and network of pores. Although some of these properties can be estimated by SAXS, it does not provide information on the accessibility of pores (i.e. it does not differentiate between open and closed/buried pores), as well as their adsorption characteristics for a specific adsorbent. Therefore, studying adsorption of a probe molecule on to porous nanomaterials, called porosimetry, has become a powerful technique. There are generally two types of porosimetry techniques: mercury intrusion porosimetry and gas adsorption. The former is more suitable for measuring macroscopic length scales ($\gg$100 nm) and it operates under high pressures, which can at times crush or break down nanomaterials. As such, gas adsorption is commonly used for probing the porosity of nanomaterials which contain (mainly) internal and some external porosity, in the range of a few nm or less.

Porosity at such length scales is classified into three categories [13]:
- *Macro*pores with pore widths >50 nm.
- *Meso*pores with pore widths between 2 nm and 50 nm.
- *Micro*pores with pore widths <2 nm.

It is generally the micro- and meso-porosity that are of interest in many cases that involve the adsorption of small molecules (e.g. reactants or pollutants) as a key step in the given application. Gas adsorption can be used in manometric (or volumetric) or gravimetric modes, with the former being the most common. Typically, the probe molecules used include N_2, Ar, and Kr, with nitrogen being the most common. In a typical measurement, the probe gas is dosed in a chamber containing the sample and

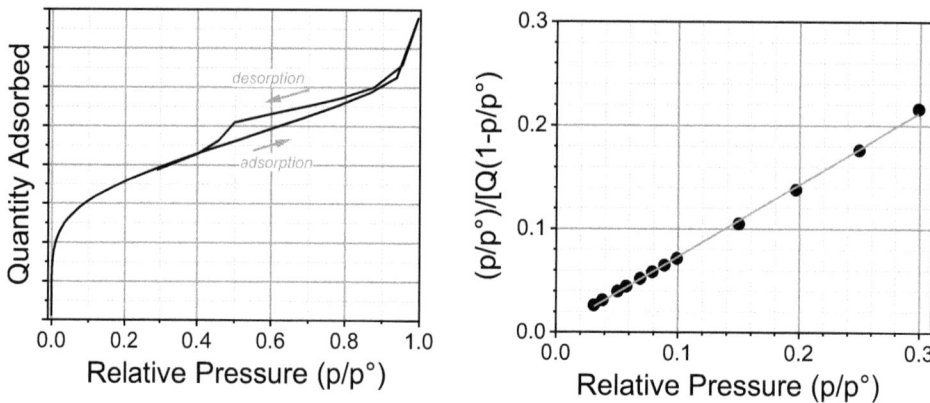

Figure 3.12. Typical gas adsorption data collected (left) and analysed using a BET method (right) to determine the BET surface area.

allowed to equilibrate at a constant temperature and partial pressure (p/p^o, where p is the equilibrium pressure and p^o is the saturation vapour pressure). The amount of gas adsorbed is measured before increasing the partial pressure in stages until saturation vapour pressure is reached. A similar procedure is used in desorption mode, where the amount of gas desorbed is measured as a result of sequentially reducing the pressure. These data are used to produce an adsorption isotherm, as shown in figure 3.12. This isotherm has various features, including initial sharp adsorption due to micropores, and the presence of hysteresis due to complex pore structures (e.g. pore networks and pore blocking effects). The details of the isotherm types, their physical significance and analysis are outside the scope of this section (they are also widely described in depth elsewhere, see [13]). However, it is worth describing the method used to obtain a specific surface area (a_s) for a given sample once the amount adsorbed (Q) at a range of partial pressures is measured (as shown in the left graph in figure 3.12). The procedure developed by Brunauer–Emmett–Teller (the 'BET method'), is the most widely used method to estimate a_s, and is based on monolayer adsorption capacity (Q_m). The data from the isotherm is converted in to a BET equation as follows:

$$\frac{p/p^o}{Q\left(1 - \frac{p}{p^o}\right)} = \frac{1}{Q_m C} + \frac{C-1}{Q_m C}\left(\frac{p}{p^o}\right) \tag{3.8}$$

Using this equation in the monolayer adsorption region (typically between the p/p^o range of 0.05–0.3), a linear BET plot is created by plotting the left hand side of this equation against the partial pressure as shown in figure 3.12 (right). The straight line in this plot provides the value of the monolayer capacity (Q_m) and an indication of the monolayer adsorption energy (via C).

The BET specific surface area is then calculated using the following equation:

$$a_s = Q_m L \sigma_m / m \tag{3.9}$$

where Q_m is obtained from the BET plot, L is the Avogadro constant, m is the mass of the adsorbent sample used in the measurement, and σ_m is the molecular cross-sectional area of the probe molecules (taken as 0.162 nm^2 for N_2). A number of methods exist for determining the pore sizes and their distribution and due to the complex and material-specific nature of the analysis, readers are directed to specialised texts [13, 14].

3.6 Summary: key lessons for characterisation of nanomaterials

This chapter has given a brief overview of several different characterisation techniques used by researchers to understand and assess materials produced. It should be noted these are brief introductions, and more in-depth literature can and should be sought from the references in this chapter, and from more detailed, focused papers for those intending to specialise in any specific technique. This chapter gives an assessment of such characterisation with nanoscale materials in mind, and a selection of the most suitable methods. However, after reading this chapter and before attempting to characterise a nanomaterial, or assess/review if methods used by others to characterise a new material are adequate/suitable and valid, one should first consider what the purpose of characterisation is. On the most basic level, it is to 'see' the material: the structure, the shape and size, the homogeneity, the chemical elements present. Characterisation is also undertaken to obtain the 'feel' of the material: to understand it in relation to the chemical and physical properties. This may require more experimental thought and design on the nanoscale than on the macroscale, as the materials themselves are not visible to the naked eye. A few key lessons to think about beyond this chapter are as follows.

- Design the experiment and analysis well. Think: what do you want to find out? What question is being asked of the material produced? Then ask: what is the right characterisation technique to find that out? Sounds obvious, but so often researchers will simply go through the motions of repeating the same analysis for everything they produce. The mistake here is that all the design and imagination goes into the synthesis of the new material, with little consideration for the design of analysis. However, the analysis is equally, if not more important, if you want to see and understand what has been made with any detail and robust rigour.
- As has already been inferred, it is essential to completely understand the technique and methodology of characterisation that is being used. Make sure to read around the techniques chosen. This is key on so many levels. It is key to the design of an experiment, to ensure that the strengths and limitations of the method are understood. Will this technique really answer the question being asked? It is key to interpreting the data: an anomalous result may be a false negative, and to an untrained/uneducated eye lead to the whole experiment being abandoned, but a basic understanding of the technique could easily show this to be an artefact of the preparation method or analysis conditions, and correction could be made, to optimise (otherwise discarded) results.

- Use complementary techniques. Never rely on just one technique and never trust others' research that does. All techniques have their strengths and limitations and as such a full interpretation of a material cannot be presented with data from only one technique. Again, this comes back to designing the analysis. For example, if the aim is to assess the size and shape of a population of nanoparticles, TEM is great for seeing the shapes, but only in a 2D projection. In addition, artefacts of drying can lead to shrinkage of some materials and incorrect sizing. 3D information can be obtained from cryo tilting, holography, and reconstructions of thin sections, and these methods are more accurate for true size. However, the limitation here is selectivity. The whole population is not considered. Scattering techniques are better in this respect, but one must understand the method to understand if the data is giving the particle size or the solvated size. What is assumed? That the particle is spherical? Can and should mathematical constants used to calculate the size be adjusted for the sample? Are the answers obtained realistic in relation to other results? And so on; e.g., can the results from both microscopy and scattering agree (considering everything that is now understood)?
- Do not rely on us (this chapter!): we all get old and time moves on. Characterisation techniques are improving at an astonishing rate. The 'go to' techniques that most people teaching used for their undergraduate and postgraduate research projects will be very old now. Unless supervisors/mentors specialise in the latest cutting edge characterisation, you should look further afield too. Look up the latest instrument for the question you want to ask, for the analysis you have designed (see above). Then look up who has that instrument and how you can access it. Do not just use what is in your lab because it is there. The chances are it will be a good characterisation method, because it is in a nanomaterials research lab, but only use it because it is the correct instrument to use, not just because everyone else in the group uses it! Use the characterisation methods that are right for the material and answer the questions you want to ask, with more than one method per question, to gain a full and robust analysis.

References

[1] Kelsall R, Hamley I W and Geoghegan M 2005 *Nanoscale Science and Technology* (New York: Wiley)
[2] Oshikane Y, Kataoka T, Okuda M, Hara S, Inoue H and Nakano M 2007 *Sci. Technol. Adv. Mater.* **8** 181
[3] Lemmer P, Gunkel M, Baddeley D, Kaufmann R, Urich A, Weiland Y, Reymann J, Müller P, Hausmann M and Cremer C 2008 *Appl. Phys.* B **93** 1
[4] Weisenburger S, Boening D, Schomburg B, Giller K, Becker S, Griesinger C and Sandoghdar V 2017 *Nat. Methods* **14** 141
[5] Duckett S and Gilbert B C 2000 *Foundations of Spectroscopy* (Oxford: Oxford University Press), 90
[6] Bird S 2016 A bioinspired approach to Data storage *PhD Thesis* University of Sheffield

[7] Galloway J 2012 Biotemplated arrays of nanomagnets using the biomineralisation protein Mms6 *PhD Thesis* University of Leeds
[8] Mei Y *et al* 2010 *Nat. Mater.* **9** 768
[9] Roach P, Farrar D and Perry C C 2006 *J. Am. Chem. Soc.* **128** 3939
[10] Figueiredo J L, Pereira M F R, Freitas M M A and Órfão J J M 1999 *Carbon* **37** 1379
[11] Pogula S D, Patwardhan S V, Perry C C, Gillespie J W, Yarlagadda S and Kiick K L 2007 *Langmuir* **23** 6677
[12] Patwardhan S V, Taori V P, Hassan M, Agashe N R, Franklin J E, Beaucage G, Mark J E and Clarson S J 2006 *Eur. Polym. J.* **42** 167
[13] Thommes M, Kaneko K, Neimark Alexander V, Olivier James P, Rodriguez-Reinoso F, Rouquerol J and Sing Kenneth S W 2015 *Pure Appl. Chem.* **87** 1051
[14] Cychosz K A, Guillet-Nicolas R, Garcia-Martinez J and Thommes M 2017 *Chem. Soc. Rev.* **46** 389

IOP Publishing

Green Nanomaterials
From bioinspired synthesis to sustainable manufacturing of inorganic nanomaterials
Siddharth V Patwardhan and Sarah S Staniland

Chapter 4

Conventional methods to prepare nanomaterials

4.1 Top-down and bottom-up methods

Existing methods for producing nanostructured materials are generally categorised as 'top-down' or 'bottom-up'. Top-down methods involve starting with a large 'piece', which is subsequently reduced and shaped to a desired size and form (figure 4.1). Conceptually, this is analogous to wood carving, where a final sculpture or artwork is made by carefully removing unwanted parts using specialised tools (e.g. a chisel). In the context of nanomaterials, the starting point for top-down methods is a bulk material, which can be fragmented to smaller sizes using etching or milling, until a desired size, or the limit of the method, is reached. Sometimes, these are also called destructive methods, because they rely on material removal. On the other hand, bottom-up techniques start with smaller objects, which are assembled together into a desired form using specialised mechanisms (figure 4.1). This is similar to using bricks to build a building or creating Lego structures. Hence they can be viewed as constructive methods, as they rely on material addition. In this case, the particle formation pathways all the way from atoms/molecules to cluster and nanoscale structures are controlled by modulating the reaction chemistry, nucleation and growth.

Table 4.1 lists selected examples of the most commonly used top-down and bottom-up methods for producing nanomaterials—some of these are currently commercially used while others are in the development stages. Figure 4.2 shows various processing techniques as a function of the sizes of objects/features they can produce, and the progress made with time. In the early days, *small* objects/features were made using precision machining techniques. Soon after, new methods such as lithography started to emerge, which pushed the boundaries towards sub-micron features. Simultaneously, bottom-up chemical approaches were developed. As both approaches moved into the nanoscale (1–100 nm), techniques using a combination of the two approaches were discovered. The sections below provide an overview of selected methods from both top-down and bottom-up approaches, with

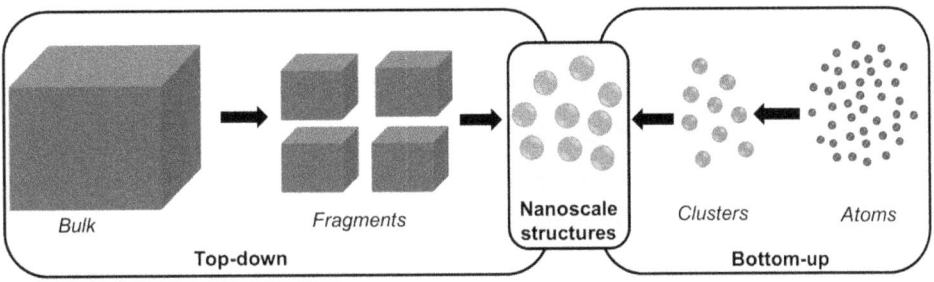

Figure 4.1. Schematic representation of top-down and bottom-up approaches to producing nanomaterials.

Table 4.1. Selected examples of the most common methods for manufacturing and synthesising nanomaterials. Adapted from [1].

Top-down methods	Bottom-up methods
Lithography	*Vapour-phase methods*
Photolithography	Molecular beam epitaxy
E-beam lithography	Atomic layer deposition
Immersion lithography	Pulsed laser deposition
Lithography with particles	Sputtering
Nano-imprint lithography	Evaporation
Soft lithography	Laser ablation
	Flame synthesis
Etching	*Liquid-phase methods*
Wet etching	Precipitation
Dry etching	Sol-gel
	Solvothermal synthesis
	Sonochemical synthesis
	Microwave-assisted synthesis
	Reverse micelle
	Electrodeposition
Electrospinning	*Self-assembly methods*
Mechanical attrition/milling	Electrostatic self-assembly
	Self-assembled monolayers
	Langmuir–Blodgett films

suitable examples. The purpose is to give an idea of current and future industrial methods, rather than providing an exhaustive list and in-depth description of all available methods. In particular, we focus on methods that are most widely used for commercial production, as shown in figure 4.3.

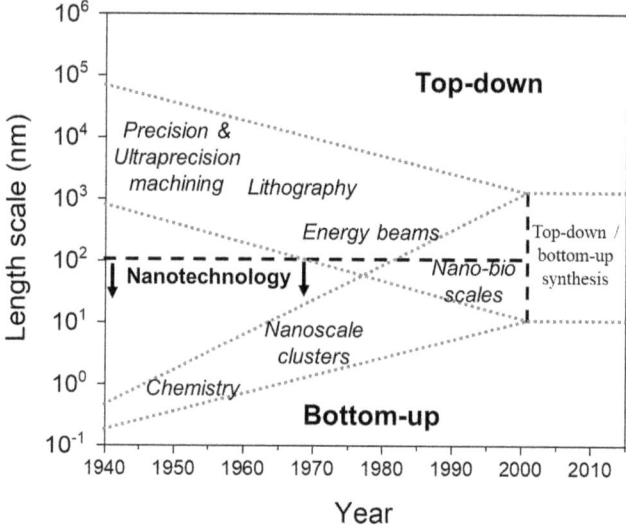

Figure 4.2. Various manufacturing techniques shown as a function of the length scales they can achieve and the progress made over time. Image adapted from [2].

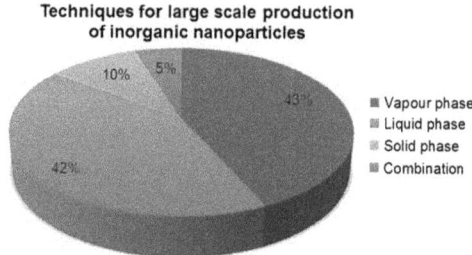

Figure 4.3. Distribution of techniques used for inorganic nanomaterials at large scales. Image adapted from [3].

4.2 Top-down methods

Top-down approaches typically start with a bulk 3D material in the form of ceramic/metal powders ($\gg\mu$m sizes) or a bulk 2D surface in the form of a pre-deposited thin film ($\approx\mu$m thick). Using various techniques listed in table 4.1, nanostructured patterns are created on the films, or the powders are reduced to smaller sizes, with varying levels of control over the process. Most commonly, in the case of films and surfaces, some form of lithography is used.

Lithography involves patterning a surface using light, ions or electrons [1]. Lithography either increases or decreases the stability/solubility of selected areas. Upon subsequent etching, the less stable/more soluble areas are removed. A mask is used to selectively remove parts of the surface, thus leaving behind a desired pattern. Although electron- and ion-lithography can create features in the nanoscale (even <10 nm), their speed is too slow for production timescales. Instead, optical (or photo-) lithography is widely used due to its fast production rates and economic operations. Research on photolithography in recent years has enabled patterning features at 10s of nm length scales [4]. Lithography is mostly limited to high value or high performance applications, such as electronic chips. It also requires specialised environments such as a vacuum and high energy source.

Mechanical attrition constitutes size reduction using physical techniques such as crushing and milling, which has been used for production of 'fine' powders since the 1970s. Traditionally, these methods are used for reducing the size of particulate solids (e.g. minerals and ores) from ≫mm to the mm–μm range, while specialised techniques can produce particles in the sub-μm range (e.g. a colloid mill) [5]. These methods have the advantage of operating under low temperatures and can be solvent-free. There is the possibility of operating in reactive or unreactive mode. In the latter, the precursors are milled to desired sizes typically using ball mills. When operated in reactive conditions, metal precursors are milled in the presence of a reactive gas (e.g. O_2) and allowed to react, thus forming nanocrystallites of metal oxides. These methods are also called mechanochemical processing, and in recent times they have been used for producing sophisticated materials such as metal–organic frameworks.

Mechanical attrition has been also used to produce new alloys and mixtures at large scales. These techniques operate under the principle of mechanical breakage of solids by imparting energy (via impact). The energy required for a given size reduction operation will be directly proportional to the increase in surface and the reduction ratio. It is clear that mechanical or physical size reduction operations consume large amounts of energy, and as the desired product sizes push beyond the submicron range, the energy required will be substantially higher. Another drawback of these methods is the limited control over particle sizes, shapes and size distribution. Monodisperse particles in the sub-micron range are seldom possible with mechanical attrition.

4.3 Bottom-up methods

In order to synthesise (nano)materials from a bottom-up approach, there are fundamental chemical principles and methods involved in building materials from atoms and molecules. These principles are described in depth elsewhere [6], and here we provide a summary of these principle reactions by focusing on the physical state of the precursors. A number of methods exist that involve the reaction in a single phase—either in solid state, solution or gas phase. One reaction proceeding in solid state is *metathesis exchange*. A metathesis exchange reaction starts with precursor salts (e.g. a metal chloride and a sulphide) producing a metal sulphide. A *single source reaction* also occurs in solid state where a precursor, typically a metal complex, is thermally converted to a desired material. Both types of method produce by-products that are solids or gases (e.g. NaCl or CO_2). Solid state methods rely on heating to provide the activation energy for the reaction and hence typically occur at high temperatures (sometimes in excess of 1000 °C).

There are a number of reactions that proceed in solution state, and these occur at lower temperatures compared to solid state reactions. The increased mobility of molecules in the liquid state helps overcome the activation energy at lower temperatures. Reactions include a thermally driven *direct reaction* between stoichiometric amounts of precursors (e.g. chemical precipitation), which is very common given its simplicity. Another commonly used solution method involves *sol–gel chemistry*, where metal alkoxide solutions in a water–alcohol mixed solvent are hydrolysed in the presence of a suitable catalyst, with further polycondensation leading to the formation of sol first and then a gel. Further, this gel is typically thermally processed to remove the solvent and complete the condensation reaction to produce an oxide material (see section 4.5). *Hydrothermal and solvothermal* syntheses are also solution phase methods, which occur at high temperatures and pressures. While hydrothermal synthesis occurs in aqueous medium, solvothermal methods use non-aqueous solvents. In the majority of the solution phase methods, given that the products coexist in the solvent with the unreacted precursors and by-products, the downstream separations become important to obtain pure materials.

Mixed-phase syntheses are also common where the reactants are in different phases. These include reactions between solids and liquids (e.g. *ion exchange* reactions between an oxide in a melt of a salt or *intercalation* of dissolved ions into a dispersed solid) and between solids and gas (e.g. *chemical vapour deposition* onto a hot solid substrate or *electrochemical* synthesis/deposition from a solution onto a substrate). Improved reactivities in a gas/liquid phase in combination with heating provides the activation energy for these types of reactions.

4.4 Nucleation and growth theory

How do atoms aggregate to form nanoparticles, is the ultimate bottom-up synthesis question. Specific for the liquid phase, but also relevant to some forms of vapour phase synthesis is the question of how particles nucleate and grow (described in section 4.5). It is important to understand how this process occurs so one can understand the size, shapes and distribution (homogeneous or polydispersed) profiles that these particle populations establish.

Nucleation is the starting point in the formation of either a new material or thermodynamic phase via the aggregation/assembly of the smallest building blocks of that material, be they atoms/ions/monomers or molecules. The nuclei are the smallest[1] clustering of these building blocks and can be considered the first step of the particle formation process. Nucleation occurs when the small units such as atoms or molecules come into close proximity and aggregate. This occurs primarily by increasing the concentration, increasing the likelihood of atoms coming into contact with one another. Once nucleation occurs and particles begin to grow, the soluble atoms are 'used up' in forming the nanoparticle, so the concentration begins to drop off until growth ceases once the atoms have reached too low a concertation to continue to grow the particles. This theory of nucleation and particle growth is over half a century old and is still a good model (figure 4.4). There are two forms of nucleation: homogeneous and heterogeneous, detailed below.

---

[1] It should be pointed out that research in this field has now identified a smaller entity still in the liquid phase, called a 'pre-nucleation cluster'.

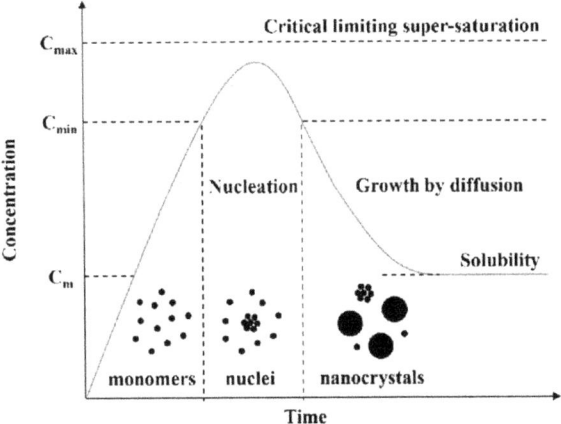

Figure 4.4. Concentration of a monomer over time, as the concentration is increased, C_m is the saturation concentration, C_{min} is the minimum concentration for particle nucleation, C_{max} = critical limiting super-saturation. Reprinted with permission from [7], copyright 1950 American Chemical Society.

4.4.1 Homogeneous nucleation

Homogeneous nucleation describes the formation of all nuclei at the same time in a bulk solution. This occurs spontaneously from a highly saturated solution or vapour. The key to homogeneous nucleation is high concentration. The nucleation is driven solely by monomer species interacting enough to aggregate. So what are the driving forces in homogeneous nucleation? We can consider this thermodynamically with the Gibbs free energy of the forming nuclei. There are two competing energies:

The volume free energy—this is negative, thus favourable and describes the energy of the bulk within the forming particle.

The solid/liquid interfacial energy—this is positive, thus unfavourable and describes the energy associated with the interaction of the surface of the forming particle with the surrounding media.

Thus the volume free energy favours the growth of the particle, while the interfacial energy favours the dissolution of the particle. The more surface area there is compared to the bulk particle, the more likely the particle will not form, as dissolution is favoured. This is why a sugar cube is slower to dissolve than smaller granular sugar. This is summarised in the following equation (4.1):

$$\Delta G = \underbrace{A_{SL}\gamma_{SL}}_{\text{Interfacial term}} + \underbrace{V_s \Delta G_V}_{\text{Volume term}} \qquad (4.1)$$

where A_{SL} is the surface area term ($= 4\pi r^2$), and V_s is the volume term $\left(= \frac{4\pi r^3}{3}\right)$.

Therefore, at small particle size (with small r) the interfacial term will dominate and the nuclei formed will redissolve back into the solution, but at a critical nucleus size

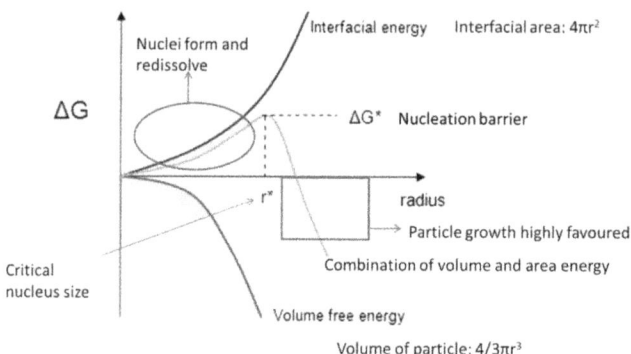

Figure 4.5. The free energy versus radius of homogeneous nucleation.

(r^*) the free volume energy term begins to dominate and the total free energy of nucleation begins to fall, quickly resulting in it being more favourable to form and grow particles (figure 4.5).

The theory shows that once the radius of the nuclei has surpassed the critical radius ($r > r^*$), the particle will grow. This also shows that the critical radius is the lower limit of the particle size. Thus if nanoparticles smaller than the critical radius are required, the surface energy must be reduced, or they will be unstable and redissolve. This can be done in a number of ways, such as using a different solvent, or by changing the surface in other ways, e.g. adding an additive to bind to the surface.

The rate of nucleation (J) is thus dependent on the thermodynamic nucleation barrier (ΔG^*) and the concentration of monomer species. This concentration is defined within the kinetic constant K, which represents the number of monomer species able to interact (per unit volume). This is the collision frequency (equation (4.2)):

$$J = K \exp(-\Delta G^*/kT). \tag{4.2}$$

The equation shows that at a given concentration and temperature, the nucleation barrier is overcome, and then all the nuclei start to grow into particles spontaneously. The higher the concentration and the lower the energy barrier, the higher the rate of spontaneous nucleation.

It should become apparent on reflection that these are difficult and artificial circumstances to create (although it can be done with a supersaturated NaCl solution) and thus homogeneous nucleation is rarer than heterogeneous nucleation based solely on increased concentration of monomers. In reality, other factors to promote aggregation apart from just concentration such as chemical reactions resulting in decreased solubility, will drive nulceation. However, the theory is very useful to understand the process of both types of nucleation.

4.4.2 Heterogeneous nucleation

Heterogeneous nucleation describes nucleation on a substrate, such as a container wall or the surface of impurity particles. It is much more common, as it does not require a saturated concentration of the crystallising species. Nucleation can occur at much lower concentration because the nucleating surface acts to lower the interfacial energy, effectively reducing the nucleation barrier. Basically, a particle forming on a surface is less exposed to the surrounding medium, has less surface atoms compared to bulk and thus has a lower interfacial energy. By this rationale, while a flat surface is good, a step offers even less exposure to the solution (two sides are contacting the substrate so in the 'bulk'), and a crack even less (figure 4.6).

This explains why you can encourage crystallization of a solution by scoring a glass flask that contains it with a spatula. This makes a rough groove in the glass to promote crystallisation in this groove.

As a simplistic model, the reduction in energy barrier can be calculated for a flat surface by multiplying the homogeneous free energy barrier by a (fractional) shape factor, as demonstrated in figure 4.7 and shown in equation (4.3).

$$\Delta G^*_{het} = S(\theta) \times \Delta G^*_{hom}$$
$$\text{Shape factor } S(\theta) = \frac{(2 - 2\cos\theta + \cos^3\theta)}{4}$$
$$= \frac{(2 + \cos\theta)(1 - \cos\theta)^2}{4} < 1 \quad (4.3)$$

($\theta = 0$ or $180°$ is homogeneous nucleation).

4.4.3 Growth

How particles grow once nucleated is highly dependent on the species concentration, the type and rate of nucleation that has occurred, and any other 'additives' that are present in the media/solution.

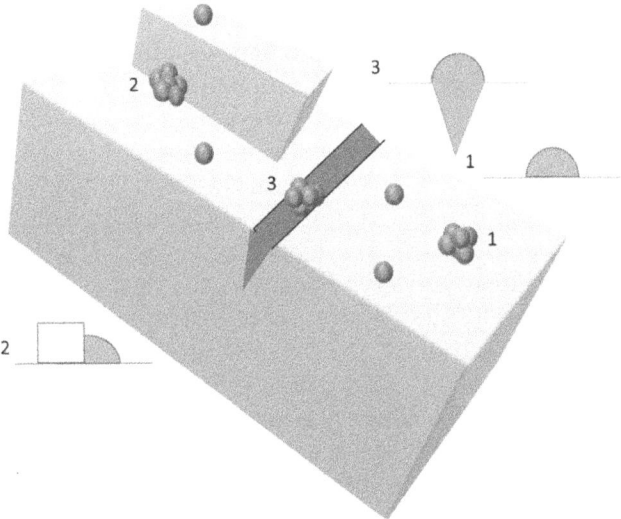

Figure 4.6. Showing heterogeneous nucleation in different surface types. (1) A flat surface, (2) a step and (3) a crack.

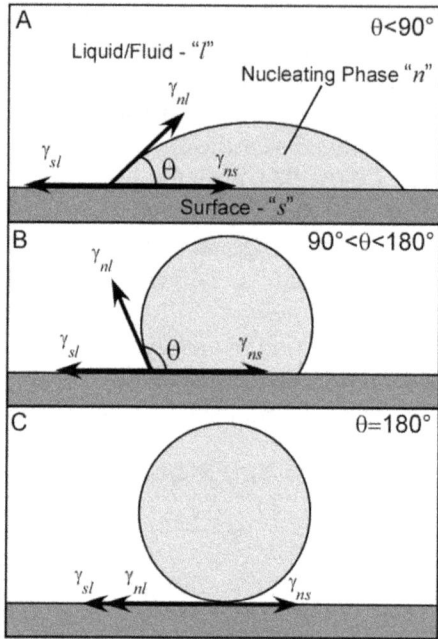

Figure 4.7. Schematic of the possible interactions of a nucleus with a surface at various contact angles according to the Young equation. (A) shows a favourable interaction at $\theta < 90°$, (B) shows an unfavourable interaction at $90° < \theta < 180°$, and (C) shows no interaction ($\theta = 180°$), which is the same as homogeneous nucleation (in the absence of a surface). Figure reproduced from [8], copyright 2012 American Chemical Society.

Nucleation defining growth: At one extreme, if a rapid, spontaneous, **homogeneous nucleation** occurs, all the particles will grow quickly, at the same time and the same rate until the concentration of the species in solution drops below a value where particle formation ceases. This will **result in homogeneous particle population of the same size**. The size will be dependent on the growth rate and concentration. If many particles are nucleated very rapidly, then the concentration will reduce rapidly, resulting in a population of many small particles, whereas a growth of less particles will result in a population with a larger particle size. The key, though, is that a homogeneous population is achieved if all the nuclei are spontaneously formed at the same time.

Heterogeneous nucleation can occur at various concentrations depending on the surface and the shape/roughness of the surface. This means that particles can nucleate at different concentrations and different times in the same solution. Thus some particles will be growing while others will still be nucleating in the same solution, leading to a **more heterogeneous size population.** A phenomenon called **Oswald-ripening** can occur which reduces the polydispersity of particle sizes somewhat. This is where the more unstable, smaller nuclei will dissolve back into the solution enabling the larger particles to grow further.

4.4.3.1 Additives restricting growth

The addition of either a soluble or insoluble additive can restrict particle growth and thus change a particle's shape/morphology. An **insoluble additive** will take the form of a **template or mould** for the particle to grow to fill or coat, depending on the template. Furthermore, crystallisation in confined spaces has an impact on the resultant material as well as its shape [9]. Several biological templates will be discussed in this context in chapter 7. Soluble additives such as small molecules can also define a particle's morphology by restricting growth of some facets over others.

A particle will nominally grow spherically at first when it is simply a cluster of atoms, as this reduces the surface energy. However, after this nucleation stage most crystals will inherit the shape defined by the crystallographic unit cell for that crystal structure. These will have defined crystal faces as facets of the particles. If an additive has a particular propensity to bind strongly to one particle face over another, the morphology of the growing crystal can be changed. A crystal face with a bound additive will have a lower surface energy, it will be more stable and thus grow more slowly than higher energy surfaces. In fact, molecules designed to stop ice forming do so by binding to the ice nuclei crystals and suppressing ice crystals from forming completely. However, if the additive binds to only one type of face only, the growth of this face is suppressed, meaning the other faces grow more rapidly and ultimately grow into extinction, leaving the additive-bound stable, slow growing face to dominate. This can be conceptually difficult to grasp, and is aided by a diagram (figure 4.8). Again this idea is expanded on with biological face-selective additives in chapter 7.

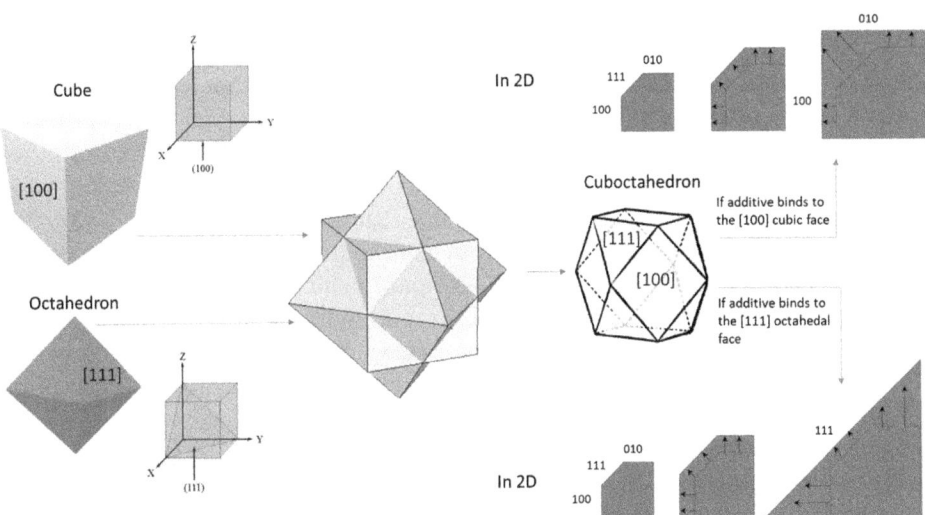

Figure 4.8. Illustration of how a cuboocahedral is formed by combining a cube and an octahedral (by cutting of the corners), and how an additive will direct the formation back to either a cubic or octahedral morphology.

4.5 Conventional bottom-up methods

This section focuses on the methods that are currently in use for large scale production of nanomaterials, while the next section is devoted to emerging methods that are showing promise for future commercial implementation. Bottom-up techniques can be broadly divided into two categories: (a) vapour phase condensation, and (b) chemical precipitation from salt solutions (table 4.1) [10]. Depending on how the energy is provided to the system or the precursor (thermally or mechanically), various sophisticated methods can branch out of these two broad categories. Examples of chemical precipitation include precipitation from an organometallic precursor or spray conversion, while examples of vapour phase methods include laser ablation, microwave plasma and spray pyrolysis. A selection of the most common methods are described below.

4.5.1 Vapour-phase method

Typically, vapour-based methods include the evaporation of metal or other precursors, followed by their deposition under vacuum on a cold surface [11]. In the case of oxides and ceramics, the chemical conversion is induced thermally, e.g. using flame process. Suitable gases can be introduced to achieve the desired composition or chemical conversion. E.g. O_2 for oxides, NH_3 for nitrides and CH_4 for carbides. This is a versatile method, which can use vapour, aerosol or fog (liquid in gas) as a feed. Flame synthesis, in particular, is widely used industrially to manufacture materials such as titania and silica despite the high temperature operations. A key advantage of this method is convenient separation of the product from the reactants and by-products, because only the product is in solid state, while other compounds are in a gas phase.

4.5.2 Solution processing

This is a simple chemical synthesis/precipitation method that is widely used to produce large quantities of nanomaterials. Aqueous solution processing starts with metal salts (e.g. chloride or nitrate) dissolved in water, followed by neutralisation, displacement and/or thermal conversion reactions to form nanoparticles, which agglomerate and precipitate. The conditions required for this method vary, but typically require low temperatures (room temperature or with slight heating to ~80 °C). The colloidal stability is a key parameter that controls the particle formation and precipitation. Typical examples of materials produced include oxides such as silica, titania, zinc oxide and manganese oxide. Although practiced widely, this method lacks the versatility to produce high quality and/or multi-component nanomaterials of desired composition.

Aqueous solution processing is also used to produce metallic nanoparticles by chemically reducing aqueous metal salt solutions. The particle sizes are controlled via the choice of the precursor, the reducing agent used, the solvent and the temperature. Addition of surfactants or capping agents is commonly used to arrest nanoparticle growth and avoid aggregation (detailed in section 4.6.5). From a commercial viewpoint, the downstream processing is challenging and costly (e.g. the separation of nanoparticles, or the removal of unreacted precursors).

4.5.3 Spray conversion

This method is used to produce multicomponent nanomaterials. This can be achieved by producing each constituent particle first and then combining them, typically via sintering. Alternatively, in spray conversion, the aqueous solutions of the precursor salts of each constituent are mixed and then spray dried in order to produce a 'composite' precursor powder. This precursor is then thermochemically converted to produce the desired multicomponent nanomaterials. Important examples include tungsten carbide and cobalt composites [10].

4.5.4 Sol–gel method

This is a highly versatile solution-based, low-temperature (~60 °C–120 °C) method useful for producing a range of multicomponent oxides and composites [11]. Sol–gel processing works on the principle that metals can react with water (hydrolysis) forming metal hydroxide colloids ('sol'), which can further react or condense to form a gel under controlled conditions. The use of catalysts for both the hydrolysis and condensation steps, the co-solvents and the precursor chemistry play crucial roles in controlling the sol–gel process. This route has been widely adopted at commercial scale in the production of high value nanomaterials such as oxides and ceramics (e.g. silica, titania, ZnO), particularly where multicomponent materials with a high level of control over their properties is required.

4.6 Emerging bottom-up methods

In recent decades, a range of new methods have been invented and developed for the bottom-up synthesis of 2D and 3D nanomaterials. A predominant common feature of these methods is the use of self-assembly of atoms, (bio)molecules or clusters composed of organic, inorganic or hybrid species. Before discussing individual methods, let us first grasp the principles behind self-assembly.

4.6.1 Principles and overview

Self-assembly can be defined as (after [12]) a spontaneous process (under the right conditions/trigger) which results in 'aggregates' of length scales larger than its components, such that there is a higher order, stable structure and/or specific properties, typically not present in its components. These special features are rendered due to the ability of the components to recognise specific binding sites, resulting in selectively positioning themselves into the final non-covalently connected structures. The self-assembled structures can be static or oscillating between different states. They find vast applications in templating nanomaterials; we will describe this further with an example later in this section.

Self-assembly is extremely important in bottom-up fabrication used for both 'soft' and 'hard' materials [12]. Components that can self-assemble, fall between the sizes that can be controlled chemically and those that can be manipulated by conventional manufacturing. Hence self-assembly offers a crucial technique to bridge top-down and bottom-up methods (described below with the example of soft lithography). As self-assembly requires components to be mobile, it typically occurs in the fluid phase.

As such, significant progress has been made in the synthesis of nanomaterials using bottom-up approaches which employ sophisticated techniques such as self-assembly, nano–bio interactions, and templated synthesis [6, 12, 13]. Self-assembly is an integral part of bottom-up synthesis approaches. It allows materials and structures to organize themselves at different length scales, using a concerted combination of weak and noncovalent chemical interactions—hydrogen bonds (water-mediated), ionic bonds (electrostatic interactions), hydrophobic interactions and van der Waals interactions—between individual molecular building blocks. Some great examples of self-assembly are biological building blocks, ranging from nucleotides and amino acids to sugars and lipids, that self-assemble to form nucleic acids, proteins, peptides, and supramolecular structures such as membranes and vesicles. From the industrial manufacturing point of view, cooperativity in the assembly of the building blocks at different length scales is an important principle. It offers the opportunity to control the material architectures at various length scales. Manipulation of molecules into functional nanomaterials using self-assembly requires a deep understanding of the following:

- The complex nature of the weak and non-covalent chemical interactions that lead to thermodynamically stable nanostructures.
- Pathways for assembling building blocks into larger and more complex nanomaterials starting from a limited number/type of precursors.
- The chemically-coded information in the molecular building blocks to obtain self-assembly (conformation and charges).

Self-assembling nanomaterials range from natural biopolymers (DNA, proteins) to small-molecular hydrogels and organic–inorganic composite materials such as metal organic frameworks (MOFs) [14]. Nanomaterials from the assembly of polypeptides and DNA with inorganic clusters/particles have been produced by several research groups for an array of applications, ranging from gene therapy to electronics [15].

Accordingly, small molecules exhibiting molecular recognition such as oligopeptides or PEG can be used to take advantage of their similar gelation behaviour to biomolecules. These materials are particularly useful in biomedical applications due to their often biocompatible nature, leading to exploration for drug delivery or as medical adhesives. In addition to their greater abundance than biomolecules, these gels are made at ambient conditions with only mild initiation required (e.g. oxidative conditions [16] or enzymatic decomposition [17]), therefore representing a low environmental and energetic cost of manufacture.

With this background information on principles involved in bottom-up methods, in particular the role of self-assembly, let us consider selected examples that have been of enormous interest for fabricating/producing nanomaterials. As these methods are of emerging nature, there are limited commercial products from these methods, however, they have great potential in future applications.

4.6.2 Soft lithography

Soft lithography uses a combination of traditional lithography, self-assembly and chemical synthesis to produce patterned features, mainly on surfaces, creating two-dimensional materials. Typically, these features can span the sub-micron to nano-scales and therefore they are of considerable interest, as traditional lithography cannot reach these length scales without difficulties. Further, soft lithography enables the patterning of features with chemical (charged, acidic or basic), physical (wetting, hydrophobic or hydrophilic), biological (proteins or cells), electrical (conductivity), etc functionalities [6].

Briefly, soft lithography works by using traditional photolithography to create a master stamp (or a mask or mould) made up of a polymer (polydimethylsiloxane (PDMS)) [18]. This stamp is then used to pattern the surface of interest via patterning with chemical 'ink' (e.g. micro-contact printing). Subsequent sequential chemical treatments can create surfaces patterned with the desired functionality. The following procedure describes the entire process step by step, and this is also summarised in figure 4.9.

1. The first step in soft lithography is to create a *mask* with the designed pattern. This can be achieved by a suitable computer aided design (CAD) package and traditional lithography.
2. Next, a photoresist layer is created on a substrate such as silicon wafer.
3. The photoresist layer is then selectively exposed to UV light via the *mask* to cure the exposed areas.
4. Unexposed (or exposed) areas are then washed away by chemical treatments to create a '*master*' for the desired stamp.
5. This master is surface treated with fluorosilane to reduce its stickiness and allow easy removal of the stamp once casted.
6. The master is then filled with PDMS pre-polymer, cured and then released to obtain a *stamp* with the pre-defined pattern.
7. The patterns on the PDMS stamp are then coated with a chemical ink (typically alkane thiols) via simply dipping the stamp in the ink.
8. The substrate desired to be patterned is pre-coated with gold (and titanium) to allow the ink to be transferred via thiol–gold bonds.
9. The ink is then *printed* on the substrate to create a chemically patterned surface. One common way of achieving this is micro-contact printing, which means simply stamping the chemical ink carefully on the substrate.
10. The parts of the substrate that remain uncoated with the ink are *etched* away by washing with a cyanide and hydrofluoric acid.
11. Finally, subsequent *finishing* treatments can be performed to produce a patterned surface with the desired chemical and physical properties. These treatments can include depositing additional chemicals or activating the surface for its use.

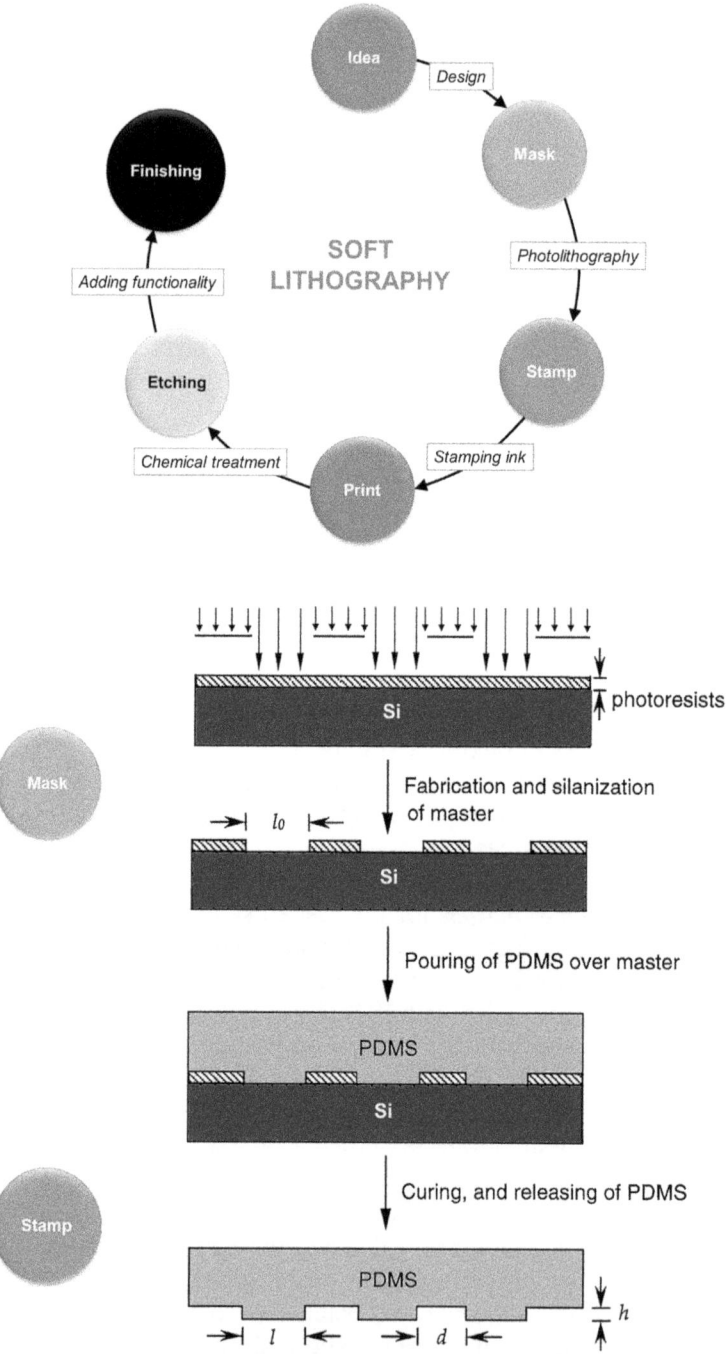

Figure 4.9. Overview of the steps involved in soft lithography (top); schematic showing how a PDMS master stamp is made (middle) and used for patterning surfaces (bottom). Adapted with permission from [6, 18, 19], copyright Royal Society of Chemistry, John Wiley & Sons and Annual Reviews. Typical values for h, d, l_0 and l are 0.2 ± 20, 0.5 ± 200, 0.99l and 0.5 ± 200 μm, respectively. Stages shown in the top image are clearly identified in the rest of the figure. (The figure is continued on the next page.)

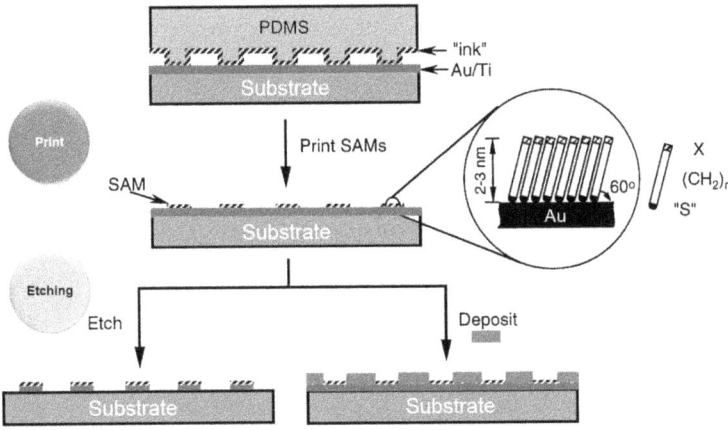

Figure 4.9. (Continued.)

The selection of the ink (from alkane thiols of a variety of functional groups to even cells) and chemical treatments in steps 10 and 11 (including the growth of nanoparticles/crystals) enable the generation of a range of surface functionalities. A clear novelty of soft lithography is how elegantly it combines top-down methods (steps 1–6 and 10) with bottom-up methods (steps 7–9 and 11).

4.6.3 Dip-pen nanolithography

An obvious limitation of soft lithography is its inability to reach nanoscales readily when patterning surfaces (see typical dimensions in figure 4.9). In order to address this challenge, nanoscale quills can be used to directly 'write', instead of PDMS stamps, to transfer/create patterns onto substrates. A tip of a scanning probe microscope, which is ~10 nm or smaller in size, is an ideal nano-quill. This has resulted in a new method called dip-pen lithography, which is a further extension of soft lithography. Dip-pen nanolithography (DPN) works by depositing molecules from their solution droplet adhering to the tip or the pen directly on the substrate, as shown in figure 4.10. DPN has opened up the possibility of patterning down to 10 nm. A diverse range of inks have been used for DPN, such as molecules, enzymes, nanoparticles, reactive precursors and viruses, for patterning various substrates.

4.6.4 Layer-by-layer self-assembly

This method is driven by electrostatic self-assembly between oppositely charged species containing multiple ions. These polyions can be deposited one layer at a time on charged surfaces. Although a single electrostatic 'bond' is much weaker than a covalent bond, the formation of multiple electrostatic attractions provides robustness to these films/layers. A typical procedure for performing layer-by-layer (LbL) assembly includes starting with priming the surface of a substrate to render electrostatic charge. This is typically achieved by chemical treatment with reactive

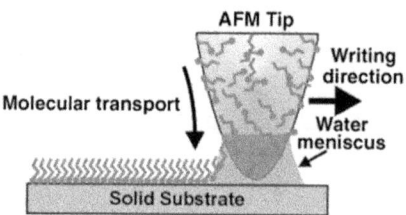

Figure 4.10. An overview of DPN. Image reproduced with permission from [20], copyright 2004 John Wiley & Sons. AFM stands for atomic force microscope, a form of scanning probe microscope.

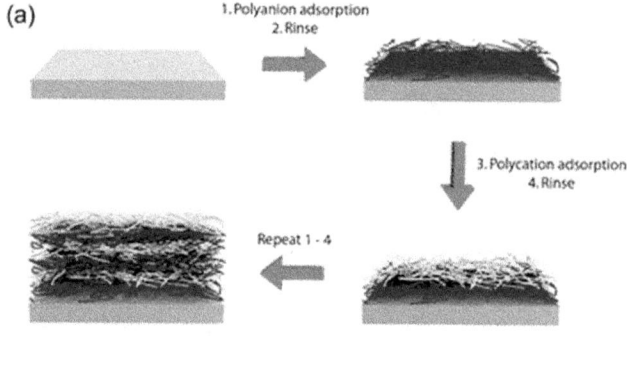

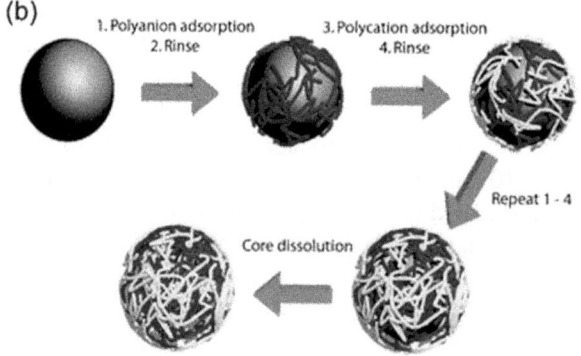

Figure 4.11. Overview of LbL assembly. Image taken from [21], reproduced by permission of The Royal Society of Chemistry.

molecules such as silanes and thiols. Next, the surface is treated with a polyion, rinsed and then coated with the next polyion of opposite charge, and rinsed (see figure 4.11(a)). This process can be repeated multiple times to obtain multilayers. An interesting feature is that the substrate does not need to be a flat surface; LbL works even on particles, as shown in figure 4.11(b).

The versatility of this method can be appreciated further once we consider the possibility that the polyion species could be polymers, proteins, peptides, polynucleic

acids, viruses, colloids or functional nanoparticles. By manipulating the conditions (pH, ionic strength and solvents) and selecting the desired chemistry of the polyions (composition and structure), the LbL growth and properties of the final structures can be controlled. Some examples of this method include the fabrication of:
- conducting layers, by using electroactive nanoparticles or polymers,
- stimuli responsive surfaces, by exploiting the swelling behaviour of polymers under certain conditions to adsorb or release active components (e.g. pollutant removal or drug delivery),
- porous multilayers, by selectively dissolving colloids from a colloid–polyelectrolyte assembly and then using the pores to host enzymes.

4.6.5 Solution synthesis of nanoparticles

A number of solution-based synthesis methods exist for nanoparticles. They use the principles of direct reaction in combination with solvothermal or hydrothermal synthesis. When performing a direct reaction, for example in chemical precipitation, a suitable precursor (generally a salt) is dissolved in water and 'neutralised' to induce the precipitation of particles. Although widely used in industry, this method offers limited control over particle sizes, their polydispersity and structure (which is important for crystalline materials). Given the importance of such methods for producing nanomaterials in bulk, in order to address these issues, an 'arrested nucleation and growth' technique has been developed for the controlled synthesis of nanoparticles. This method uses an organic capping agent to rapidly trap nuclei and allow controlled slow growth of these nuclei into nanoparticles. The capping agents also help improve the colloidal stability of the nanoparticles thus formed. This method has been extensively used in the past two to three decades for the synthesis of a range of metallic nanoparticles, which is reviewed elsewhere [6, 22, 23].

With the examples of two semiconductor nanoparticles (CdSe and CdS), we describe this synthesis method. Typically, a capping agent is selected such that it can also act as a solvent. Trioctylphosphine oxide (TOPO) is a commonly used capping agent cum solvent in the synthesis of CdSe nanoparticles. The synthesis occurs by heating dimethyl cadmium in TOPO to 300 °C, followed by rapid addition of tributylphosphine selenide. Over several hours of growth period, TOPO capped nanoparticles are obtained by washing steps and antisolvent precipitation using acetone. In such synthesis, high temperature is required to generate high quality crystallinity, while the solvent/capping agents control the particle sizes and polydispersity. CdS nanoparticles are also synthesised in a similar fashion where oleylamine is used instead as a capping agent solvent. Sulphur is directly added to a pre-heated solution of $CdCl_2$ in oleylamine [22]. The growth occurs slowly over up to 6 h at temperatures of 140 °C–160 °C, to form nanoparticles, which can be again isolated with antisolvent precipitation using ethanol. These are some of the early and pioneering examples of controlled nanoparticle syntheses in solution. There have been a very wide range of investigations reported where modifications are made to these methods. Modifications include the substitution of the capping agents with more intricate chemistries and/or compatibilities with solvents of choice. A range of surfactants and polymers have been used as capping agents. Other modifications include a two-step synthesis where in the first step a metal-complex is prepared, which is then reacted in the second stage, allowing further control over the particle formation.

4.6.6 Templated synthesis

Aside from using self-assembly to produce a material itself, self-assembling templates are often used to impart long-range order in higher-volume materials (mostly inorganic). Templated synthesis enables the formation of nanoporous materials. These can be viewed as nano-sponges where the pores can be of the sizes of small molecules (<2 nm, called micropores) or larger (between 2–50 nm, called mesopores, and >50 nm, called macropores). Zeolites (microporous aluminosilicate networks), also called molecular sieves due to their ability to 'filter' molecules based on their sizes, are often templated around single molecules such as small alkylammonium ions. Materials templated around self-assembling surfactant molecules have begun to introduce mesopores and more complex structures into materials. The templates self-assemble into micelles or liquid-crystals, leading to porous materials with either larger porosities than are available through simple molecular templating alone, or lamellar/interconnected structures [12].

These features are best illustrated with an example of one of the most studied materials—mesoporous silicas (see figure 4.12), although we note that similar strategies have been applied to many other nanoporous materials. In a typical

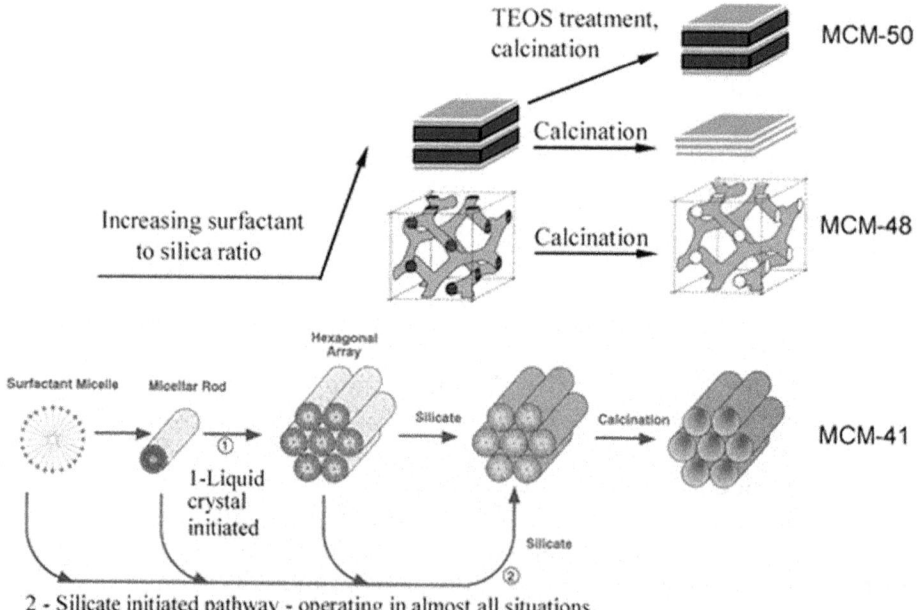

Figure 4.12. Synthesis of porous materials templated around self-assembling surfactants. Image reproduced by permission of The Royal Society of Chemistry from [24].

synthesis, cetyltrimethylammonium bromide [$(C_{16}H_{33})(CH_3)_3N^+Br^-$] is used as the template. Depending on the template concentration, ionic strength of the system, the solvents used and the temperature, the template can self-assemble into spherical micelles, micellar rods or lamellae; these rods can further pack in to hexagonal or cubic arrangements [24]. When treated with the inorganic precursor, these assemblies can template a framework around them, leading to porous materials upon template removal (typically by calcination or solvent reflux) [25]. An interesting point to note is that contrary to earlier beliefs, the surfactant assembly is affected by the presence of inorganic precursors and hence the templated synthesis includes complex cooperative assembly between the template and the inorganic species, making it more interesting [26].

4.7 Summary: key lessons about conventional routes to nanomaterials

In this chapter, we have learnt that there are a number of methods for producing nanomaterials, and that they are broadly classed into top-down (or destructive) and bottom-up (constructive) methods. Bottom-up methods are becoming more popular due to the various advantages they offer, including the ability to tailor the composition, size and structure of nanomaterials. Generally speaking, they intrinsically produce less waste. The key features in bottom-up methods are understanding and applying nucleation and growth principles. The utilisation of self-assembly (e.g. templated synthesis) and the combined use of top-down and bottom-up strategies (e.g. soft lithography) is creating valuable emerging methods for nanomaterials synthesis, which have great potential for less wasteful commercial implementation.

References

[1] Sengul H, Theis T L and Ghosh S 2008 *J. Ind. Ecol.* **12** 329
[2] The Royal Society and The Royal Academy of Engineering 2004 Nanoscience and nanotechnologies: opportunities and uncertainties, Joint report https://royalsociety.org/~/media/Royal_Society_Content/policy/publications/2004/9693.pdf
[3] Charitidis C A, Georgiou P, Koklioti M A, Trompeta A-F and Markakis V 2014 *Manuf. Rev.* **1** 11
[4] Neisser M and Wurm S 2015 *Adv. Opt. Technol.* **4** 235
[5] Backhurst J R, Harker J H, Richardson J F and Coulson J M 2002 *Coulson and Richardson's Chemical Engineering* 5th edn (Oxford: Butterworth-Heinemann)
[6] Arsenault A C, Ozin G A and Cademartiri L 2009 *Nanochemistry: A Chemical Approach to Nanomaterials* 2nd edn (Cambridge: Royal Society of Chemistry)
[7] LaMer V K and Dinegar R H 1950 Theory, production and mechanism of formation of monodispersed hydrosols *J. Am. Chem. Soc.* **72** 4847–54
[8] Hamilton B D, Ha J-M, Hillmyer M A and Ward M D 2012 *Acc. Chem. Res.* **45** 414
[9] Wang Y-W, Meldrum F C and Christenson H K 2013 Confinement leads to control over calcium sulfate polymorph *Adv. Funct. Mater.* **10** 6
[10] Kear B H and Skandan G 2000 *Ullmann's Encyclopedia of Industrial Chemistry* (Weinheim: Wiley)

[11] Scherer G W and Brinker C J 1990 *Sol-gel Science* (New York: Academic)
[12] Overton T, Weller M T, Rourke J and Armstrong F A 2014 *Inorganic Chemistry* (Oxford: Oxford University Press)
[13] Service R F, Szuromi P and Uppenbrink J 2002 *Science* **295** 2395
[14] Loweth C J, Caldwell W B, Peng X G, Alivisatos A P and Schultz P G 1999 *Angew Chem. Int. Ed.* **38** 1808
Mirkin C A, Letsinger R L, Mucic R C and Storhoff J J 1996 *Nature* **382** 607
Paul Alivisatos A, Johnsson K P, Peng X, Wilson T E, Loweth C J, Bruchez M P and Schultz P G 1996 *Nature* **382** 609
[15] Niemeyer C M 2001 *Angew Chem. Int. Ed.* **40** 4128
Aida T, Meijer E W and Stupp S I 2012 *Science* **335** 813
Zhang S G 2003 *Nat. Biotechnol.* **21** 1171
Zhang Z L, Horsch M A, Lamm M H and Glotzer S C 2003 *Nano Lett.* **3** 1341
[16] Lee B P, Dalsin J L and Messersmith P B 2002 *Biomacromolecules* **3** 1038
[17] Sadownik J W, Leckie J and Ulijn R V 2011 *Chem. Commun.* **47** 728
[18] Xia Y N and Whitesides G M 1998 *Angew Chem. Int. Ed.* **37** 550
[19] Xia Y N and Whitesides G M 1998 *Annu. Rev. Mater. Sci.* **28** 153
[20] Ginger D S, Zhang H and Mirkin C A 2004 *Angew Chem. Int. Ed.* **43** 30
[21] Quinn J F, Johnston A P R, Such G K, Zelikin A N and Caruso F 2007 *Chem. Soc. Rev.* **36** 707
[22] Joo J, Na H B, Yu T, Yu J H, Kim Y W, Wu F, Zhang J Z and Hyeon T 2003 *J. Am. Chem. Soc.* **125** 11100
[23] Daniel M C and Astruc D 2004 *Chem. Rev.* **104** 293
[24] Kresge C T and Roth W J 2013 *Chem. Soc. Rev.* **42** 3663
[25] Patarin J 2004 *Angew Chem. Int. Ed.* **43** 3878
[26] Huo Q, Margolese D I, Ciesla U, Demuth D G, Feng P, Gier T E, Sieger P, Firouzi A and Chmelka B F 1994 *Chem. Mater.* **6** 1176

Section III

From biominerals to green nanomaterials

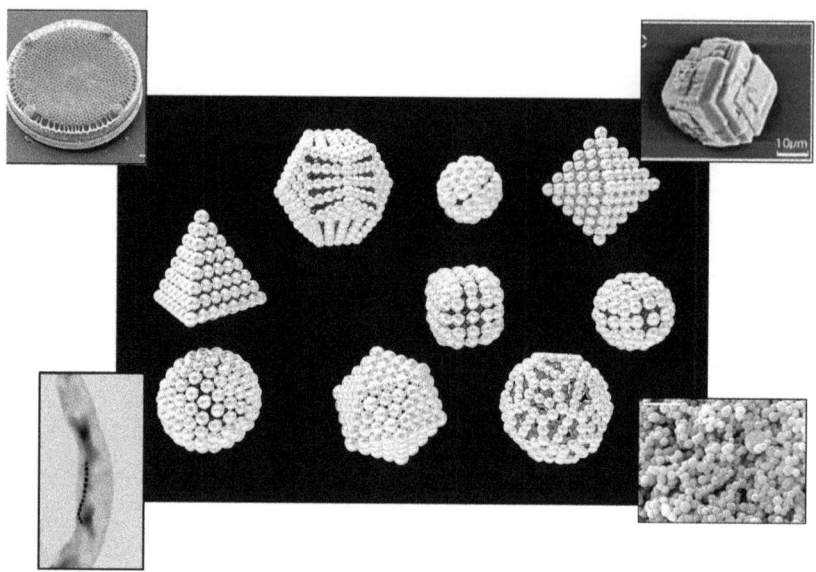

Image courtesy of Kateryna Kon/Shutterstock.

The previous section has highlighted the need for and advantages of nanomaterials, as well as demonstrating current production methods for such materials. The first section described the principles of green chemistry. This section begins by bringing together these two concepts and evaluating existing methods using green principles. Key challenges and areas of improvement are identified, and the use of biology as a source for inspiration is proposed. From here the production of inorganic materials within biology at various scales with incredible precision is evaluated in the next chapter, to gain insight into and a better understanding of the concepts and tools that Nature utilises. The final chapter of this section highlights the importance of learning from biomineralisation in order to discover new and green routes to sophisticated nanomaterials, by providing a platform for designing synthetic methods that offer the control and sophistication seen in biomineralisation yet without the need for native biological components or machineries. The key rules and insights obtained from biomineralisation are systematically identified, with examples and lessons learnt explored to develop green bioinspired routes to make nanomaterials. Moreover, the key challenges faced in scaling up and making these routes commercially viable are discussed, as these are better solved at the discovery phase.

Chapter 5

Green chemistry for nanomaterials

In this chapter, we aim to combine our knowledge from previous chapters on green chemistry principles and nanomaterials synthesis. We will evaluate how green existing nanomaterials synthesis/manufacturing methods are, and understand the reasons behind them. With selected examples of nanomaterials, the analysis of sustainability is illustrated in this chapter. It forms a platform for the remaining book by identifying ways to discover green routes to nanomaterials.

5.1 Sustainability of nanomaterials production

There are two major challenges with nanomaterials—safety and sustainability. The toxicity of nanomaterials has been widely investigated as discussed in chapter 2, however, the environmental burden from their production is largely ignored [1]. As described in chapter 4, at present, nanomaterials are manufactured using top-down (lithography, milling and etching) or bottom-up (vapour deposition, sol–gel, precipitation, pyrolysis, solvothermal) approaches [2, 3].

The success of a process relies on controlling the key specifications for a desired nanomaterial. These include nanoscale features such as particle size, particle size distribution, the degree of agglomeration and purity. At the commercial stage, the cost of the nanomaterial production also plays a key role. If the cost of the nano-product is more than traditional or 'current-best' materials, then the nanomaterials need to perform significantly better, thus providing an added value for the additional cost.

Bottom-up approaches such as hydrothermal or sol–gel synthesis are promising, but they suffer from many problems leading to high costs, an extremely adverse environmental impact (hazardous waste and toxicity problems) and unsustainable production. Indeed, the environmental impact and sustainability analysis performed using E-factor (waste-to-product ratio) revealed that the current nanomaterials production methods are up to 1000 times more wasteful when compared to the production of pharmaceuticals and fine chemicals (see table 5.1) [4]. Despite being

Table 5.1. A comparison of wastefulness of various manufacturing sectors. Adapted from [4].

Sector	Waste: product ratio (E-factor)
Bulk chemicals	<1 to 5
Fine chemicals	5–50
Pharmaceuticals	25 to >100
Nanomaterials	100–100 000

highly wasteful, and hence expensive (as described below), these approaches predominate the current manufacturing processes for nanomaterials. Significant progress has been made in the synthesis of nanomaterials using bottom-up approaches which employ sophisticated techniques such as self-assembly and nano–bio interactions. However, these recent developments still suffer from multi-step, laborious methods leading to uneconomical scale-up and/or high wastefulness.

In addition to existing nanomaterials, new nanomaterials are being invented (e.g. quantum dots, metal organic frameworks, carbon nanotubes), but due to a lack of sustainable manufacturing methods, these materials are difficult to scale-up, and uneconomical to commercialise [4, 5]. Indeed, a recent article in *Chemistry World* clearly highlighted these scale-up issues with new materials or processes [6].

5.2 Reasons behind unsustainability

Specific issues with existing methods for manufacturing nanomaterials are shown in figure 5.1, and are also listed below and described with suitable examples [3].

1. **High purity requirements**: Ultrapure precursors, reagents and solvents are typically needed, which demand high energy for purifying reagents and stake into scarce resources (e.g. potable water). Further, due to extremely low tolerance for contamination, the recycling of solvents and unreacted reagents becomes difficult, if not impossible.
2. **Low yields**: Although some progress has been made recently to improve reaction yields, the material utilisation efficiencies over the entire process (reaction, separation, finishing) are poor for nanomaterials, ranging between 1%–20% (typically 3%–10%), causing wastefulness; such methods are not suitable for scale-up.
3. **Repeated/sequential processing**: Many processes for nanomaterials synthesis/manufacturing require sequential steps (e.g. layer-by-layer deposition) including repeated washing for purification. Due to low efficiencies in each step, the waste produced multiplies.
4. **Toxic or hazardous chemicals**: Each of the current methods for nanomaterials either use or produce toxic or hazardous chemicals including greenhouse gases (e.g. the use of metal halides or alkoxides as precursors in bottom-up methods, and emission of PH_3 gas from thin film deposition).

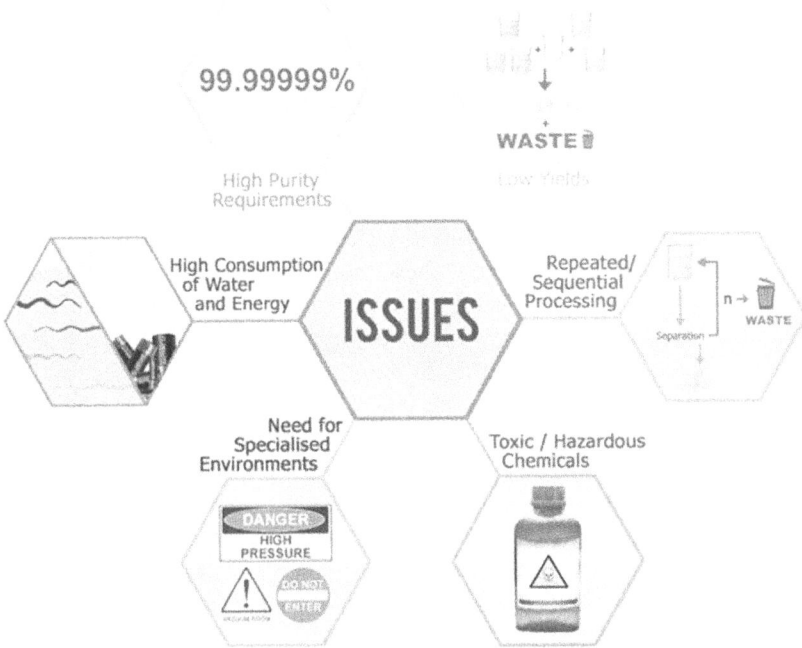

Figure 5.1. Summary of the issues associated with current manufacturing methods for nanomaterials. Courtesy of Yung Hei Tung.

5. **Need for specialised environments**: Specialised environments such as ultrahigh vacuum (<5 torr), or high temperatures to melt metal salts (>1000 °C), are typically required for nanomaterials production, which create an additional cost and energy penalty.
6. **High consumption of water and energy**: Most of the issues listed above dictate, directly or indirectly, high consumption of water and energy. It has been estimated that to create a 1 cm^2 nano-device, 10–60 L of water and 1 kWh energy are required, which create an enormous burden on the environment.

5.3 Evaluation of sustainability for selected methods

5.3.1 E-factors for solution methods

In order to illustrate these issues, we provide examples of nanomaterials synthesis. The E-factor results shown in table 5.1 consider three types of nanomaterial—carbon nanotubes, titania and gold nanoparticles—and selected methods for each [4]. It was found that for titania produced using an industrial hydrolysis–calcination method (a hybrid of sol–gel and vapour-phase synthesis) resulted in an E-factor of 17800. This means that to produce 1 kg of titania using this method, nearly 18 tonnes (yes, tonnes!) of waste is generated. For solution synthesis of gold nanoparticles, the E-factor was between 3320 and 99400 for chemical processing, depending on the capping agent and solvent used. A key finding was that downstream processing such as purification increased the waste produced significantly. This means that when reporting and evaluating the sustainability of a method for nanomaterials synthesis, we should quantify the washing steps and avoid phrases such as 'washed with copious amounts of solvent', 'rinsed until no further change in conductivity was recorded', or 'the antisolvent was added until precipitation occurred'. In the case of semiconductor nanoparticle synthesis (section 4.6.5), simply noting the temperatures required (>140 °C), the duration of the synthesis (>6 h) and the chemicals involved, it is clear that these methods are unlikely to be regarded as sustainable. Attempts to improve resource efficiency sometimes result in two-step methods, where in the first step a metal complex is prepared, which lowers the temperature and/or time to synthesise nanomaterials in the second step. As explained in chapter 1 (equations (1.8), (1.11) and (1.12)), such modified methods may indeed be more unsustainable as they overlook the environmental burden of the first step and only focus on the improvements in the second step. An example of this is the synthesis of CdSe quantum dots using a 'single-source precursor' Cd $(S_2CNMeHex)_2$ containing a complex of Cd and Se [7]. Heating this precursor in tri-n-octylphosphine (TOP) and tri-n-octylphosphine (TOPO) produced high quality CdSe nanoparticles, perhaps with properties better than a common solution synthesis explained in section 4.6.5. However, when one looks at the synthesis of the single-source precursor, which uses carbon diselenide, the authors noted 'carbon diselenide is a very toxic, evil-smelling liquid and extreme caution must be observed during preparation and manipulation.' This clearly means that the first step of this two-step synthesis is highly unsustainable, thereby perhaps not improving the green credentials of the overall method.

5.3.2 How green is soft lithography?

Soft lithography, described in chapter 4, has been very popular in recent decades and is considered to be a revolutionary method for producing nanomaterials. It is important to reiterate that soft lithography has truly extended surface patterning to length scales that were not readily accessible using traditional lithography or other methods. Therefore, let us consider its sustainability in order to illustrate the points made above. At this point, we note that we could not find a single published article related to the life cycle or environmental analysis of soft lithography [1]. We will therefore have to use qualitative information in order to assess this method. The first thing that strikes one when looking at the soft lithography process (see section 4.6.2 and figure 4.9), is how rather lengthy the procedure is. It requires more than 11 steps, which means that the method is bound to suffer from the issues faced by sequential and repeated processing. Next, consider steps 4, 10 and 11, which pertain to some sort of washing or etching. As soft lithography is based on sequential addition and selective removal of layers, these steps are bound to produce relatively large amounts of waste (compared to the actual material un-washed/retained as a pattern), resulting in high E-factor. Further, it is remarkable to note the variety of chemicals required for this method; most of these are extremely hazardous and require extreme care during and specialised environments for handling. These include fluorosilane, used in step 5, and cyanide and hydrofluoric acid used in step 10. Although not generally reported, it is likely that the yield or production rates (patterned area produced per time) are low, particularly when it takes up to a day to create one layer thick patterns. Combining all these points together, it appears that soft lithography would score very low on sustainability or green chemistry ratings. To balance this, it is very important to consider this environmental cost or burden against the benefits offered, to make an educated prediction of its value. This is important because sometimes the ability to pattern at desired length scales offers a wider benefit over a longer term.

5.3.3 Templated synthesis: surely sustainable?

Similar to the methods discussed above, in the case of templated materials, e.g. mesoporous silica considered in section 4.6.6, there remain several issues with these materials, both in terms of the common requirement for hydrothermal synthesis conditions and for the need to remove the template materials prior to use. Although it has been shown to be feasible to modify the synthetic conditions such that hydrothermal methods are not necessary, mild template removal methods are still in their infancy. Therefore, there has been significant research into either templates that are designed for non-destructive removal, or more efficient removal methods—either by weakening the surface–template interaction or increasing the solubility of surfactant in another way such that irradiation, oxidative conditions, or solvent reflux—are capable of purifying the material [8]. Despite the improvements found by these studies, issues with the quality of materials resulting from these sustainability-minded compromises remain, either in terms of residual template contamination or inferior properties. As a result, no such methods have progressed to larger than lab-scale syntheses.

5.4 Adopting green chemistry for nanomaterials

It seems that most methods that we considered revolutionary have given little attention to their sustainability. Despite their preparation being very wasteful and environmentally damaging, most attention aside from material performance has only focused on the toxicity of nanomaterials [9]. However, existing methods for manufacturing nanomaterials suffer from issues pertaining to sustainability. These issues dictate high consumption of water and energy, creating an enormous burden on the environment and result in unsustainable manufacturing, yet the environmental burden of the synthesis/manufacturing process remains largely ignored [1, 10]. These factors clearly stress the urgent need for developing fundamentally new production methods for nanomaterials that are green and sustainable.

Adopting the principles of green chemistry and engineering has the potential to develop greener routes (see figure 5.2) [1]. A solution suggested to overcome the sustainability issues is to adopt syntheses procedures that reduce waste and that operate at room temperature and in aqueous solutions, which in turn can lower the cost of production [9]. A recent white paper produced by the Royal Society of

	Emerging strategies for greener routes for nanomaterials				
Twelve principles of green chemistry	Design for waste reduction	Design for process safety	Design for materials efficiency	Design for energy efficiency	Design of bespoke nanomaterials
1. Prevent waste	✓		✓		
2. Maximise atom economy			✓		
3. Less hazardous chemical synthesis		✓			
4. Safer chemicals and products		✓			✓
5. Safer solvent/ reaction media	✓	✓			
6. Increase energy efficiency				✓	
7. Use renewable feedstoks	✓		✓		
8. Reduce derivatives	✓		✓	✓	
9. Use catalysts	✓	✓	✓	✓	✓
10. Design for degradation/ end of life	✓		✓		✓
11. Real-time monitoring and process control	✓	✓	✓	✓	✓
12. Inherently safer chemistry		✓			

Figure 5.2. Illustration of how the 12 principles of green chemistry can be used to design new strategies for sustainable methods for the synthesis and manufacturing of nanomaterials. This figure has been adapted from [13] with permission from The Royal Society of Chemistry.

Chemistry has identified 'green materials and processes' as a priority area for future research, in order to deliver new efficient large-scale industrial processes [11].

A World Technology Evaluation Center report [2] on research directions in nanotechnology explicitly recommended that achieving green manufacturing by 2020 is the 'holy grail'. This report suggested that future research should focus on emulating natural designs to develop scalable processes for manufacturing nanomaterials. Further, it has been highlighted that for bioinspired synthesis, future research is required to investigate scale-up, advanced processing and precise control of nanomaterials properties; their implementation into industrial manufacturing will lower the costs and broaden the applicability of nanomaterials [13]. As such, the next chapters will cover how biology makes nanomaterials (chapter 6), and what can be learnt from this, as well as the emerging rules for developing bioinspired syntheses (chapter 7). This will be followed by two case studies. Given that we will be focusing on and discussing a range of biological and biochemical concepts, in the next section we aim to introduce the relevant terminology as a simple guide, while for details, we will refer to specialised texts.

5.5 Biological and biochemical terminology and methods

5.5.1 Molecular biology component

5.5.1.1 Amino acids, proteins and enzymes

Some of the key components of biological machinery are proteins and nucleic acids. Proteins are composed of amino acids (AAs), which are also called residues. AAs have a common structure, as shown in figure 5.3 (top). There are 20 naturally

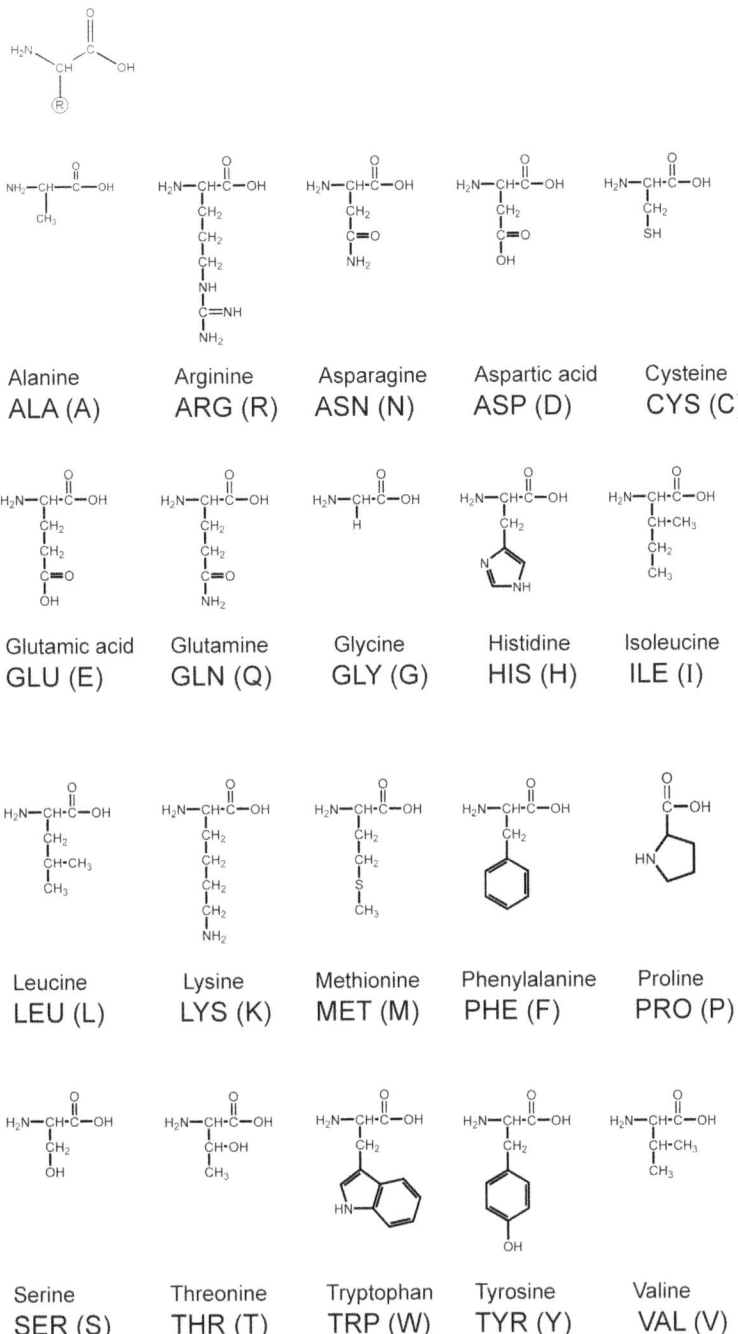

Figure 5.3. Top: generic structure of amino acids with R shown as the side/functional group. Bottom: a list of amino acids, their structures, and their three-letter and one-letter abbreviations. Image adapted from [15] with permission from The Royal Society of Chemistry.

occurring AAs, which differ from each other in the side group (shown as *R* in figure 5.3) [14]. AAs are abbreviated to either a three letter code, or a one letter code, as shown in the list in figure 5.3. For example, the AA leucine is denoted as L or LEU. The chemistry of the side group *R* provides specific properties to a given AA. These groups can be categorised as hydrophobic (aliphatic, cyclic or aromatic) or hydrophilic (acidic, basic or uncharged). Histidine, methionine and cysteine are special kinds of AAs, due to the presence of either a heterocyclic side group (histidine) or the presence of sulphur.

The amine group ($-NH_2$) from one AA reacts with the carboxylic hydroxyl group ($-OH$) from another AA to form a peptide bond. Further addition of AAs results in larger molecules (or polymers), known as proteins. The sequence of AAs in a protein, called their primary structure, ultimately determines their properties and function. Proteins fold into higher order structures as shown in figure 5.4. The secondary structure of a protein arises from the intramolecular assembly of a protein molecule, and commonly results in α-helix, β-sheet, random coil, or β-turns configurations. Non-covalent weaker bonds such as hydrogen bonding and van der Waal's forces are generally responsible for helping form these structures. Within a single protein molecule, a number of such secondary structures are common; their arrangement with respect to each other gives a protein its tertiary structure. In some cases, a number of protein molecules assemble together via intermolecular bonding to create a quaternary structure.

The structure of a protein is very important as it controls the protein function, ranging from its solubility to its recognition by other cellular components. Their ability to recognise certain targets is key in their function, an ability also exploited for controlling nanomaterials recognition as discussed in the next few chapters. Denaturation, breaking protein structure, leads to loss of function and in some cases to detrimental effects. In order to form a stable structure, there needs to be a large (> approx. 30) number of AAs. In the case of peptides, which have a much lower number of AAs, hierarchical structures are not readily possible. However, specific peptide sequences can be designed to self-assemble into supramolecular structures.

Enzymes are a special class of proteins which possess catalytic activities towards certain reactions. Enzymes are highly specific towards reactants (also called substrates) and highly selective in their function. The enzyme's structure and its catalytic 'pocket' renders its function. The pocket provides a selective site for the substrate to bind and react, the change from the substrate to the product results in it being misfit in the pocket and leads to its release.

5.5.1.2 Nucleic acids

Deoxyribonucleic acid (DNA) and ribonucleic acid (RNA) are polymers of nucleotides. There are four types of nucleotides—adenine (A), thymine (T), guanine (G), and cytosine (C)—and their sequence carries genetic information (note that these single letter notations are confusing combined with those used for AAs). As shown in figure 5.5, a nucleotide consists of a sugar and a phosphate (not shown for brevity), which form the backbone of nucleic acids. Nucleotides also contain a base

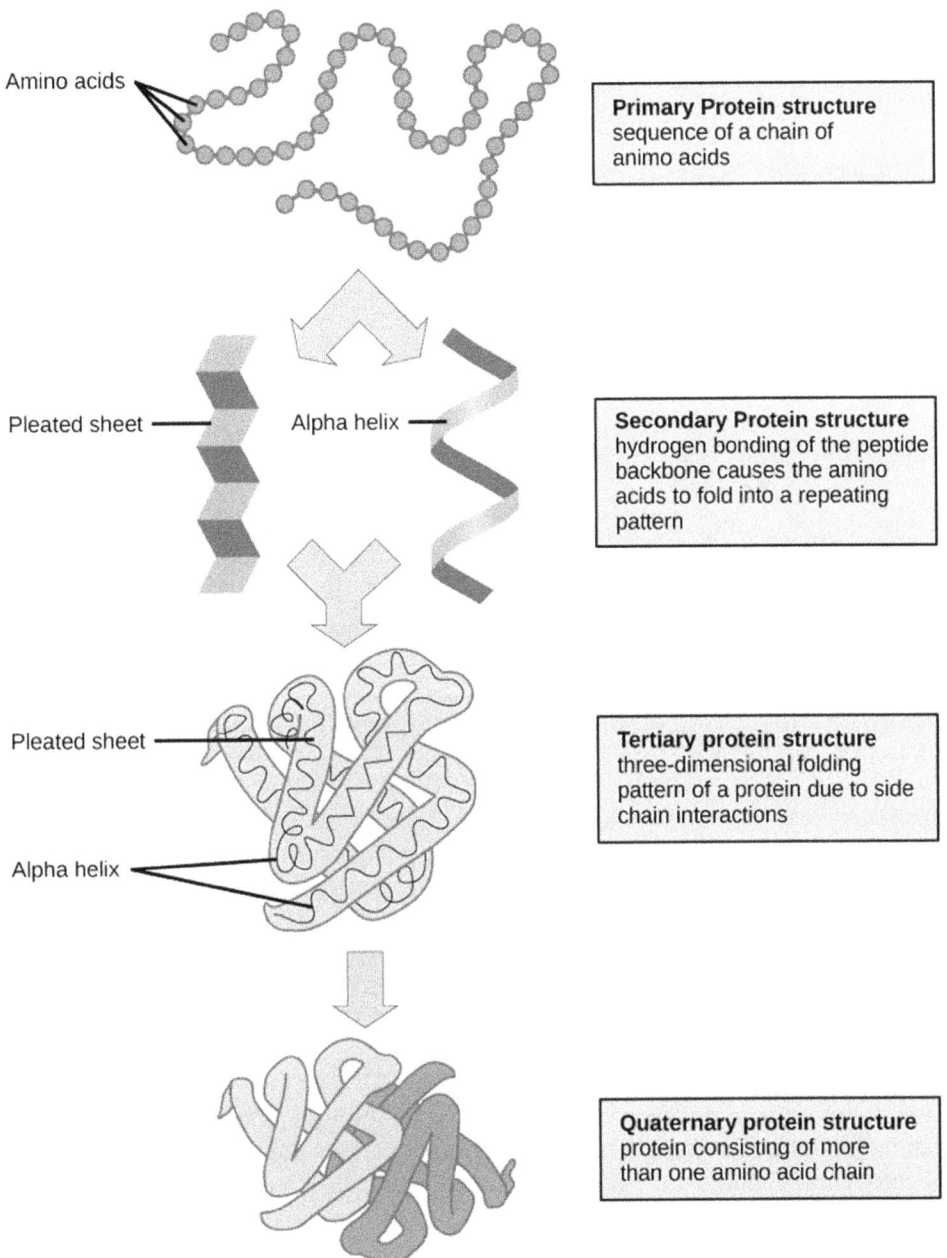

Figure 5.4. Schematic depiction of different levels of protein structure as (a) primary, (b) secondary, (c) tertiary and (d) quaternary. Source: CNX OpenStax—Wikimedia.

(see figure 5.5); there are four types of base, resulting in four types of nucleotides. Hydrogen bonding between A and T, and G and C results in the helical structure of DNA. The sequence of the nucleotides consists of the genetic code, which is

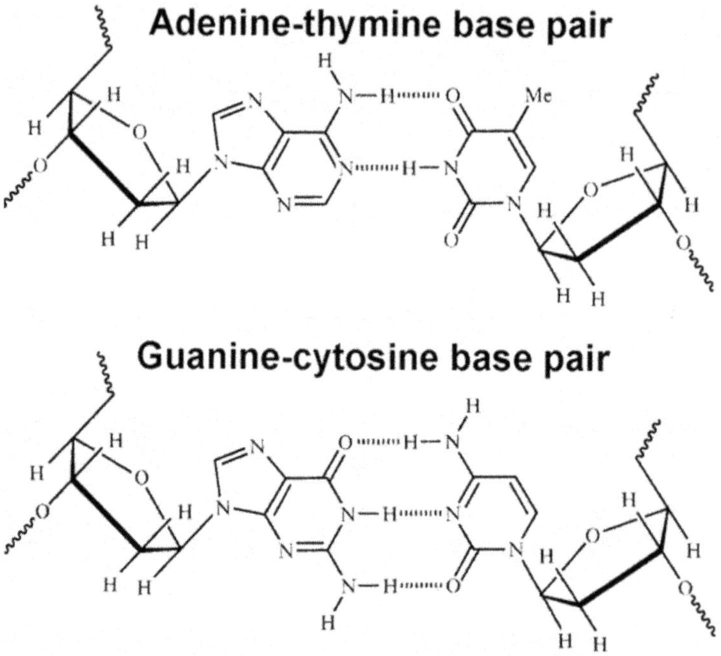

Figure 5.5. Complimentary pairs in DNA with each adenine (A) base perfectly matching a thymine (T) base to link (top), and guanine (G) and cytosine (C) match (bottom) permitting hydrogen bonding. Image taken from [17], copyright 2007 John Wiley & Sons.

'translated' into AAs and then proteins. A group of three nucleotides form a codon, which codes for a single AA. The DNA code is 'read' or transcribed by RNA, which is like DNA except that it is single stranded, and the thymine base is de-methylated (missing the CH_3), so is called uracil (U) instead of (T). When transcription needs to take place the DNA unwinds and opens up, so RNA forms the complementary sequence to the DNA. This is then fed into the ribosome, which is the cell machinery for synthesising proteins. In this way, the DNA sequence codes for a protein sequence. For example, singled stranded RNA codons UCC, GUG and CCA will translate to the AAs serine, valine and proline (S-V-P). It should be noted that as there are more combinations of 3 of the 4 bases than the 20 natural AAs, several codons will make the same AA. There are other codons that dictate at which point the sequence should start and end being read (they could be considered as punctuation codon).

5.5.1.3 Other biomolecules
There are many other types of biomolecules that proteins can produce that are key to biological functions. Those relevant to this book include phospholipids and polysaccharides, among others. Phospholipids are surfactant-like molecules with a hydrophilic phosphate head group with two aliphatic long chains, while polysaccharides are polymers composed of sugar molecules as monomers, e.g. starch and cellulose. Both of these are used for compartmentalisation, among other things, and

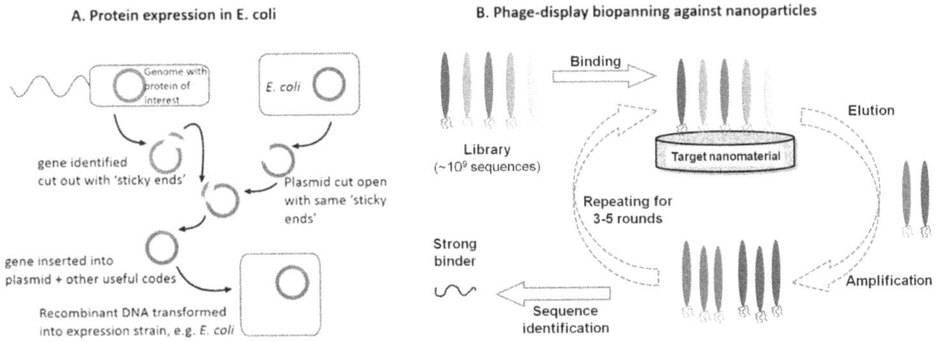

Figure 5.6. Schematics of molecular biological techniques. (A) Protein expression in a bacterial cell such as *E. coli* (reproduced with permission from [18]), and (B) phage display biopanning against nanoparticles.

are discussed specifically with respect to biomineralisation in chapter 6 (sections 6.4.2.1 and 6.4.2.2).

5.5.2 Molecular biological techniques

5.5.2.1 Protein expression in another host

Microbiological techniques have developed methods of using host cells (such as bacteria or yeast) to produce large amounts of protein from a different source. This uses the fact that many cells not only have a genome, but also have smaller loops of DNA code called plasmids. These can be easily extracted from a bacteria, manipulated and reinserted. As such the DNA code of a protein of interest can be inserted into a plasmid. This plasmid can be specially designed to contain protein purification tags and a marker gene (e.g. encode a protein that will change the colour of the cell) as well. The plasmid can then be inserted into a host cell (most commonly *E. coli* bacteria). The bacteria will then produce (or 'express') the protein of choice when it is induced, which is again dictated by the function designed into the plasmid (figure 5.6(A)). Using this method synthetic protein can be designed and expressed in a host organism such as *E. coli*.

5.5.2.2 Protein mutagenesis

Naturally occurring genes can be modified chemically or biochemically to result in a new (or slightly modified) sequence. This will result in translation of a protein with a modified sequence. This can be achieved in a number of ways: 1) disrupting a key base will then read the whole section incorrectly, as there will be a 'read frame shift'(shift how the bases are grouped into 3 so read completely different AA). 2) Whole gene or whole protein sequences can be 'cut out', and silent DNA 'pasted' back in. This genetic engineering strategy can be used to produce DNA which in turn will produce proteins of mutated sequences, or cells with specific proteins missing, known as knock-out mutants. Knock-out mutagenesis is a very important strategy for understanding which proteins are critical to which functions in a cell. In this technique, a knock-out mutant (denoted ΔgeneX) is a cell with that particular gene X missing, and thus it will not produce the corresponding protein X. The cells can then be studied to see if they are different to the native cells (or wild type, as they are often called).

5.5.2.3 Biopanning

A combinatorial approach can be used to identify novel peptides and proteins that strongly bind a target surface/material (also known as 'biopanning'). In this approach, a library of millions to tens of billions of peptides with random sequences is created. Biopanning is basically a fishing exercise. The library is exposed to a target material and the sequences that stick and bind best are separated with the material target while everything else is washed away (figure 5.6(B)). In the context of this book the targets are inorganic nanomaterials, and this is a relatively new concept as the technique was conceived more for biological targets to develop artificial antibodies. Each sequence is attached to an outer surface exposed protein of either a cell (cell display) or bacteriophage (phage display) in the library. Using a cell or phage in this way is the crux of the biopanning technique, as both the protein and the DNA which encodes it are physically linked together. This means that the recovered best candidate protein or peptides can be isolated through pH elution, and the cells or phage amplified, and thus more protein or peptide can be obtained. The process is repeated several times to evolve the strongest binding candidates that can outcompete everything else (figure 5.6(B)). Upon a number of rounds of binding, elution and amplification, tightly bound peptides or proteins can then be sequenced and expressed and their binding motifs studied.

5.6 Summary: key lessons about sustainability nanomaterials production

Based on environmental assessment of methods used for nanomaterials synthesis, we have seen that most existing and emerging methods are unsustainable. The reasons behind this include large amounts of waste produced due to low yield, sequential processing, high energy demands and the need for specialised reagents/environments. This highlights the need for a change in our perceptions and objectivity when it comes to assessing and progressing a new method for nanomaterials. We identified that biologically inspired methods have the potential to design green methods. To this end we have explored the central dogma and basic components of molecular biology and the tool kit it offers for biologically inspired nanomaterial synthesis. In the following chapters, we look for inspiration in biological mineral formation to identify rules and strategies for inventing green methods.

References

[1] Hutchison J E 2008 *ACS Nano* **2** 395
[2] Roco M C, Mirkin C A and Hersam M C 2010 Nanotechnology research directions for societal needs in 2020: retrospective and outlook *Panel Report* (Lancaster, PA: WTEC) https://www.nano.gov/sites/default/files/pub_resource/wtec_nano2_report.pdf

[3] Sengul H, Theis T L and Ghosh S 2008 *J. Ind. Ecol.* **12** 329
[4] Eckelman M J, Zimmerman J B and Anastas P T 2008 *J. Ind. Ecol.* **12** 316
[5] Dahl J A, Maddux B L S and Hutchison J E 2007 *Chem. Rev.* **107** 2228
[6] Warner J 2016 *Chemistry World* (April) p 35 https://www.chemistryworld.com/opinion/toxicity-is-a-hazardous-waste/9605.article#/
[7] Azad Malik M, Revaprasadu N and O'Brien P 2001 *Chem. Mater.* **13** 913
[8] Calugay R J, Miyashita H, Okamura Y and Matsunaga T 2003 *FEMS Microbiol. Lett.* **218** 371
[9] Patarin J 2004 *Angew Chem. Int. Ed.* **43** 3878
[10] Murphy C J 2008 *J. Mater. Chem.* **18** 2173
[11] Romero-Franco M, Godwin H A, Bilal M and Cohen Y 2017 *Beilstein J. Nanotech.* **8** 989
[12] The Royal Society of Chemistry 2011 A sustainable global society. Available at: https://www.csj.jp/international/cs3-whitepaper2010_summary.pdf
[13] Gilbertson L M, Zimmerman J B, Plata D L, Hutchison J E and Anastas P T 2015 *Chem. Soc. Rev.* **44** 5758
[14] Patwardhan S V, Manning J R H and Chiacchia M 2018 *Curr. Opin. Green Sustain. Chem.* **12** 110–6
[15] Berg J M, Tymoczko J L and Stryer L 2002 *Biochemistry* 5th edn (New York: Freeman)
[16] Patwardhan S V, Patwardhan G and Perry C C 2007 *J. Mater. Chem.* **17** 2875
[17] Currie H A, Patwardhan S V, Perry C C, Roach P and Shirtcliffe N J 2006 Natural and artificial hybrid biomaterials *Hybrid Materials* ed G Kickelbick (New York: Wiley) p 255
[18] Galloway J 2012 Biotemplated arrays of nanomagnets using the biomineralisation protein Mms6 *PhD Thesis* University of Leeds
[19] Rawlings A E, Bramble J P, Tang A A S, Somner L A, Monnington A E, Cooke D J, McPherson M J, Tomlinson D C and Staniland S S 2015 *Chem. Sci.* **6** 5586

IOP Publishing

Green Nanomaterials
From bioinspired synthesis to sustainable manufacturing of inorganic nanomaterials
Siddharth V Patwardhan and Sarah S Staniland

Chapter 6

Biomineralisation: how Nature makes nanomaterials

6.1 Introduction

Biomineralisation is the controlled formation of solid inorganic materials by biological organisms [1–3]. The very idea of biomineralisation may seem counter-intuitive to most. Inorganic minerals, like a stone, are inherently non-living, so the idea that these are intrinsically involved in life may seem strange at first. But then, consider the form of your own body. The structure and motility of your body is down to you having a calcium mineral skeleton (we are vertebrates). Other creatures have external skeletons (exoskeletal) like molluscs (they are invertebrates) that use their calcium mineral shells for protection.

From just these two examples it is clear how useful it is for biological organisms to form hard minerals, which serve a structural/mechanical and protective function. Indeed, the birth of biomineralisation is a hot topic in evolutionary biology and geology, as it appears that the rapid increase in biomineralisation and skeletal organisms occurred at the same time as the 'Cambrian explosion' between 525 and 510 million years ago. The Cambrian explosion[1] describes a time when the evolution of new species and genetic diversity rapidly increased over this time. While the reason for the rapid Cambrian evolutionary period is not yet fully understood [4] and also occurs at the same time as significant climate and sea chemistry change and instability, it is clear that the introduction of hard materials into biology introduces much greater diversity in the functions that organisms can have. For example, the earliest biominerals took the form of hard pointed spines that covered worm-like creatures, as well as very early shells on molluscs [5, 6] that act as protection within

---

[1] While it is known as the 'Cambrian explosion', researchers in the area now believe it was more of a rapid increase in diversification that should be considered in the broader context, and not an 'explosion' or 'big bang' from nowhere.

the rocky ocean environment. Fossils later showed the predator 'attack' functions of biominerals, and this went hand in hand with the development of the need to diversify to protect against predators using hard materials to form protective shells. Thus biomineralisation could be enabling diversification through a predator–prey dynamic, being a key factoring in rapid evolutionary diversification over time.

Biomineralisation occurs across the full range of biological kingdoms, from bacteria to humans, producing a range of minerals and biomineral/organic matrix hybrids which we will explore in this chapter. Furthermore we will see that Nature utilises and optimises the chemistry of inorganic material formation discussed in the last chapter, and is able to take advantage of chemical spatial definition (compartmentalisation), organic structural surfaces and chemistry at the surface, as well as strategically positioned organic functional groups, to enable the formation of complex and highly specific mineral forms and structures, all under ambient environmental temperatures and conditions. The production of superior materials under environmental conditions is a key attraction to understanding and utilising this process, as ambient reaction conditions carry huge significance for industry. While these minerals span all length scales (figure 6.1), at the end of this chapter we summarise the lessons that biology can teach us to help us adapt our current chemical synthesis of nanomaterials to more biological green synthesis of nanomaterials.

Traditionally, biomineralisation is broadly classified as follows:

- **Biologically controlled mineralisation (BCM) (figure 6.2)**: Where minerals are synthesised for a specific biological function. Biomineralisation in this class is strictly under biological (genetic) control. We will examine BCMs in this chapter.
- **Biologically induced mineralisation (BIM) (figure 6.2)**: Where minerals are precipitated due only to the environmental conditions the organism has imposed, usually as an insoluble bi-product (to a biological redox reaction, for example). These biominerals may be useful, detrimental or benign to the organisms producing them (e.g. kidney stones), and may be internal or external to the organism. As these are not precisely controlled, there will be

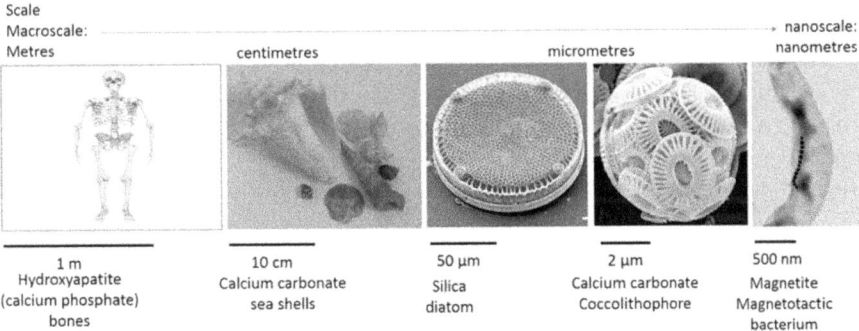

Figure 6.1. A range of biominerals at various scales, from the macro to the nanoscale. Images courtesy of Alex Mit/Shutterstock/Mary Ann Tiffany and [8].

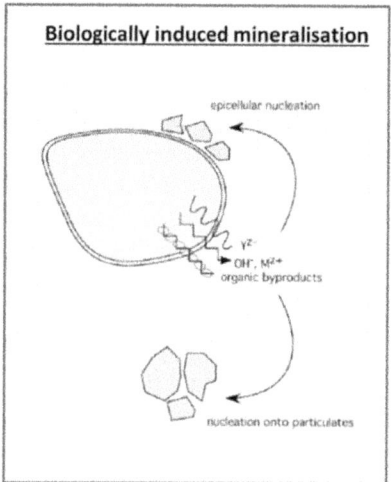 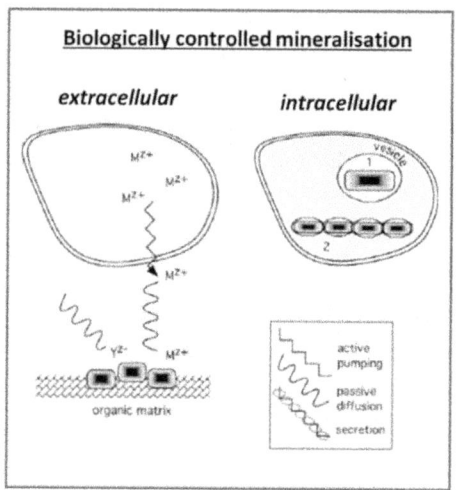

Figure 6.2. Schematic of difference between biologically induced mineralisation compared to biologically controlled mineralisation. Adapted from [2].

no further discussion of these processes in this book, and all subsequent use of the term biomineralisation will refer to BCM only.

While the biomineralisation process varies extensively, generally it involves the uptake of specific elements from the environment by an organism and their deliberate formation/incorporation into a solid mineral structure under precise biological control *in vivo*. Organisms typically accumulate the mineral precursors from their respective environments, extracted from water, soil or food. These precursors can be metal ions (e.g. ions of Ca, Fe and Mg), or even small molecular complexes ($Si(OH)_4$). Organisms are able to transport these precursors from the environment into their cell or cellular matrix. From here the precursors may be stored, then transported to the site of biomineralisation where they are concentrated to form the biominerals. It is the tight genetic control over the process that produces species-specific, ornate unique biomineral structures. Biomaterials are rarely 'pure', as they are deliberately chemically non-stoichiometric and can contain a range of dopants to enhance the functional properties. Most biominerals also have a biological organic material component that can be a molecular or larger organic matrix structure to form a hybrid composite material. Biominerals have very specific and often remarkable properties, unlike those of abiotic component materials. For example, bone has the capacity to regenerate, has massive mechanical strength, is relatively light and has some flexibility, in complete contrast to inorganic calcium phosphate. Such impressive properties evolve over time, in order for a biomineral to perfectly perform their key functions within organisms, e.g. why bone is an ideal structural/mechanical material for its purpose. The origin of these properties is not just the material's composition, indeed the precise hierarchical and often highly intricate ornate structures and morphologies of the materials on multiple length scales are equally responsible for their enhanced properties. Thus, it is critical that biominerals are formed with such highly precise levels of control over composition and structure, to ensure the best function, and this is achieved to such an extent that these complex mineral composite structures are highly reproducible.

In the next section we identify some key biominerals, and explore the link between their required function and their properties and how Nature has optimised:
- the chemical composition of the mineral;
- mineral to organic biomolecules content and interaction;
- structure and morphology;

In order to achieve the specific function. The main specific functions carried out by biominerals can be classified as follows:
- mechanical/structural support;
- protection;
- cutting and grinding;
- sensors (gravity, optical);
- cation storage;
- buoyancy.

6.2 Properties and function of biomineral types

The minerals formed by organisms most commonly utilise Ca, Mg, Si, Fe, and to a lesser degree Ba and Sr ion, as their carbonates, oxalates, sulphates, phosphates, hydroxides and oxides minerals. Relatively rare biominerals are formed from Mn, Au, Ag, Pt, Cu, Zn, Cd and Pb deposited largely in bacteria and often as sulphides or oxides.

Over 60 different biominerals have been identified, with around 50% being calcium-based; of these more than half are calcium phosphates of varying composition. The most abundant biominerals are calcium carbonates and phosphates. Calcium minerals dominate biomineralisation to such an extent that the study of biomineralisation simply used to be called 'biocalcification'. In terms of structure, 75%–80% of biominerals are crystalline, with the remaining 20%–25% being amorphous in nature, i.e. they do not show structural regularity at atomic scales (e.g. biosilica, amorphous hydrated iron phosphate, amorphous calcium carbonate).

Many biominerals are composite organic–inorganic hybrids and are structurally hierarchical, organised from the nano- to the macroscopic length scale. The organic components of biominerals can include proteins, glycoproteins, polysaccharides and other small organic biomolecules. The organic phase occluded in biominerals may or may not be directly involved in biomineralisation.

6.2.1 Bio-calcium phosphate (hydroxyapatite): mechanical/structural support, motion, cutting/grinding

Calcium phosphate is most commonly found as hydroxyapatite of the pure formula $Ca_{10}(PO_4)_6(OH)_2$. However, it is rarely stoichiometric or pure, so a more accurate and general formula could be:

$$[Ca_{10-(v+w+x+y+z)}Sr_vMg_wNa_x(H_2O)_y[]_z]$$
$$[(PO_4)_{6-(a+b+c)}(HPO_4)_a(CO_3)_b(P_2O_7)_c]$$
$$[(OH)_{2-(p+q+r+s+t)}F_pCl_q(H_2O)_rO_s[]_t]$$

showing the full range of dopants and potential vacancies ([ ]).

Bone is the most common material made of hydroxyapatite (figure 6.3). Bone is a hybrid material, with hydroxyapatite as the mineral hard component that forms on an organic matrix which forms the flexible component mainly composed of the protein collagen. The strength and stiffness can both be increased with higher mineral content, whereas it is the toughness that dictates the optimum composition for bones. Thus the ratio of mineral to organic component varies depending on the specific organism, the age of the organism and function of the particular bone. The more collagen present, the more elastic the bone; the more mineral present, the harder it is, showing this variable hybrid composition offers tuneable mechanical properties. Some animals that rely on quick motion, but can compromise on strength (due to the fact the body is supported by a water environment), benefit from having more flexible bones with higher collagen content, e.g. fish such as salmon and mackerel (>50% collagen). However, the larger the animal, the stronger the bones need to be, to support the large body mass. Thus generally the largest animals have the highest mineral content, such as cows, elephants and whales (>75% mineral). Similarly, our bones change with age. When we are born, human bones are mainly composed of cartilage, a flexible organic material which does not fracture easily, enabling us to learn and fail (by falling) without serious consequence. The cartilage is replaced over time as bone mineralises, as our need for support increases with our increasing weight, with optimal bone composition of about 70% mineral occurring around the age of 18–25. After this time the organic component begins to

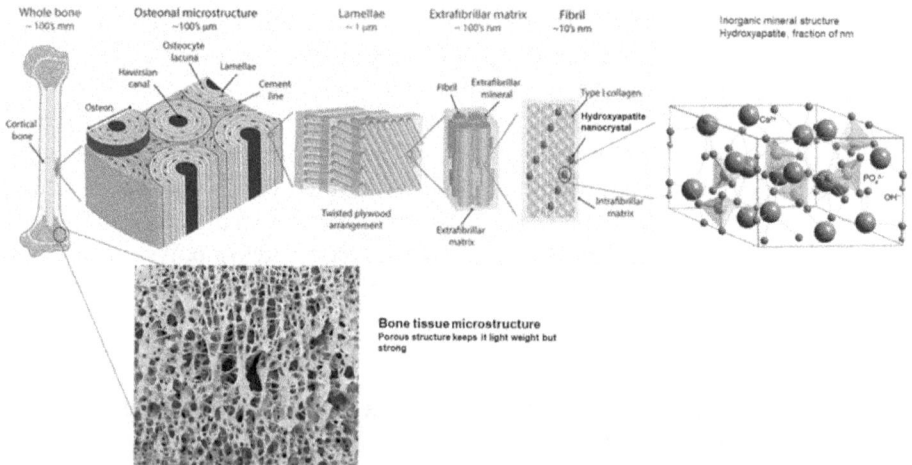

Figure 6.3. The hybrid (organic/inorganic) and hierarchical structure of bone from the atomic structure of hydroxyapatite on the right to the macroscopic structure of a whole bone on the left. Adapted from [7], courtesy of Professor Nick Greeves.

degrade slowly, leading to issues with brittle bones and increased fracturing in older people. Furthermore, bone has an intricate hierarchical arrangement composed of a fine microstructure network, which allows a blood supply, bone cells and proteins to move through the pores. In this sense bone is very much a 'living' biomineral, in that it is constantly remodelling and regenerating in response to environment factors such as physical pressure or signals from hormones. The ornate structure also contributes to enhanced properties for specific functions. Some bone needs to be extra hard, and these are dense and heavy, whereas other bone is required to be light, such as the bones of flying birds, so these are hollow. The properties of a bone can even vary within the same bone, with dense, tough joints and more porous structure in the middle.

Hydroxyapatite is also the hard mineral constituent in teeth. Like bone, teeth have a complex structure, but in simple terms the quantity of mineral increases radially outwards. The organic centre of the tooth is the 'pulp', surrounded by the 'dentine', which has a similar composition and structure to bone, with the hardest 'enamel' forming the outer coating. Enamel is required to be so hard as the purpose of teeth is to cut and grind, while no flexibility is required at all, so enamel is a high density 95% hydroxyapatite biomineral, with dopants such as F^- increasing the hardness of the enamel by reducing the solubility of the mineral, which is why we use fluoride toothpastes.

6.2.2 Bio-calcium carbonate: protection, sensor, buoyancy

Calcium carbonate ($CaCO_3$) has the largest range of biominerals, as there are several different structural forms (polymorphs) used in Nature (figure 6.4). These are

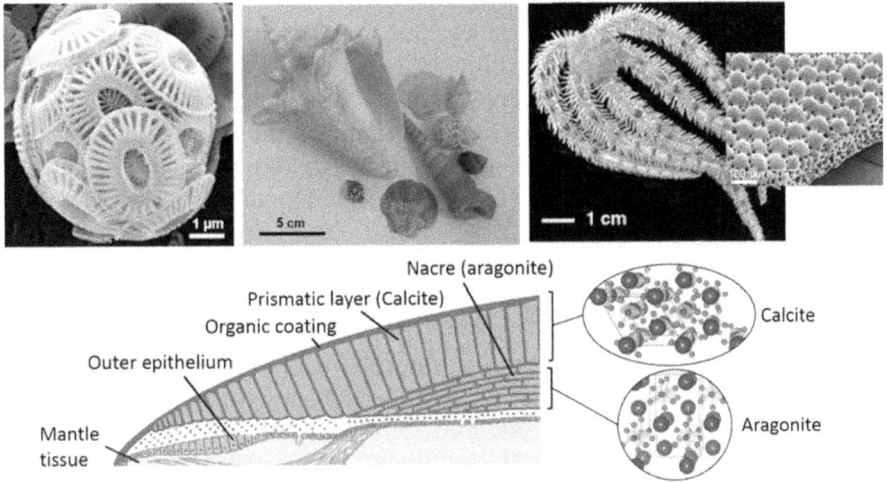

Figure 6.4. Top row: examples of calcium carbonate biomineralisation. Left: a coccolithophore [8]; centre: a collection of shells; right: a brittle star [9] (reproduced by permission of the Royal Society of Chemistry). Bottom row: schematic (adapted from [6]) of the biomineralisation of a sea shell, showing the mantle with the shell formed on top (green triangles show biomineralisation proteins), with the crystal structures of calcite and aragonite shown to the right.

mainly calcite and aragonite crystalline structures (as they are the most stable) but less stable vaterite and amorphous calcium carbonate are also found as biominerals, although are less common, with amorphous calcium carbonate, whewellite/weddellite ($CaC_2O_4 \cdot (2)H_2O$) mainly found in plants or fungi as a mineral storage mechanism for calcium.

Calcite is the most common calcium carbonate biomineral, and is the main component of sea shells, sea urchin spines, bird eggshells and the intricate plates on the very tiny unicellular coccolithophore. All of these biominerals act as a form of protection, so the key requirement is to be hard and tough. It is worth noting that as such they have a much lower organic fraction than bone (usually <5%). Again the materials can be altered with dopants, the most common instance of this is with Mg^{2+}, to make hard urchin spines even harder, for example. Some organisms use either calcite or aragonite (e.g. sponge skeletons), while many sea shells are actually made up of both [6]. Molluscs do not shed their shells, so the shells grow as the creature grows, by increasing the disc-like radius of a scallop/mussel-type shell, or increasing the size and circumference of the spiral turn of a coiled snail type shell. The mineralised shell grows from the shell mantle meaning there is very little organic material (<2%) in the shell mineral, but this biologically controls the formation of platelet aragonite nacre (mother of pearl) on the inside of the shell, and large elongated crystals of calcite form the outer prismatic layer on the exterior. Incredibly, coccolithophore biomineralise their elaborate scales internally within intracellular compartments, using a scaffold made of cellulose and proteins (called a base plate) to support crystal nucleation and growth [10]. Once fully mature, these coccoliths 'burst' out onto its surface and form an exoskeleton around the cell. Although unicellular, the white colour from the scales that coat these algae can be seen from space as huge blooms of them are visible in the ocean [8].

Calcium carbonate biominerals are used for functions other than just protection. Examples include calcite eye-lenses (for extinct trilobites and microlens arrays in brittlestars [9]) and the defensive claws of many mammals and birds. A crystal of calcite in the inner ear also acts as a gravitational sensor, with its movement activating sensory receptors. Aragonite is used as 'love darts' to aid reproduction in gastropods, and some marine molluscs and cephalopods also use aragonite as buoyancy devices.

6.2.3 Bio-silica: mechanical support, transport and protection

Another major family of biominerals is amorphous biosilica ($SiO_2 \cdot nH_2O$), which again have a mainly structural role. Their amorphous nature can be advantageous; with no set pattern or planes of crystallisation, these materials can be moulded and worked into a range of architectural lace-like structures without a loss in mechanical strength (figure 6.5). The most common example is the beautiful and intricate nano to microscale cell wall/shell structures (exoskeletons, called frustule) of the aqueous single-celled microalgae—diatoms (figures 6.5a and 6.1). Biosilica in diatoms exhibit repeating structural features of ca. 10–40 nm. The patterns are highly sophisticated over multiple length scales and are species-specific with a high degree of

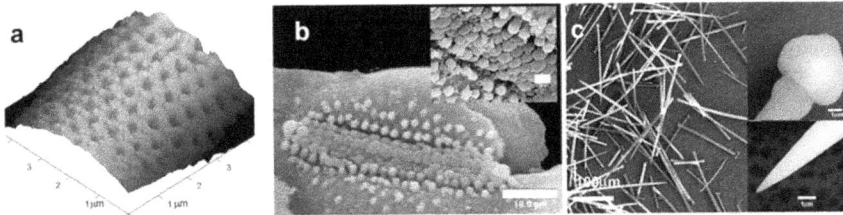

Figure 6.5. Micrographs of ornate biosilica patterns in (a) a diatom, (b) higher plants, and (c) sponge spicule. Image taken from [13], reproduced by permission of The Royal Society of Chemistry.

reproducibility [11]. These structures contain periodic pores which enable the diatom cells to communicate with other cells and the environment. It is interesting to note that diatoms form colonies by 'gluing' cells together—the pores in the silica cell walls allow communication between these cells. In addition, these pores provide a way for nutrient transport into the diatom cells. Silica makes up the biomineral skeletons for the microorganisms radiolarians and some sponges, which are known to form needle-like spicules (figure 6.5c), which are a few tens of microns in diameter and can be as long as a few millimetres. Silica is also found in plants as spines and at cell walls to provide structural support (figure 6.5b). It is the presence of silica in rice that helps the leaves adopt an upright position to maximise incident sunlight. It has been shown that without silica presence in rice plants, the leaves become *floppy* and the crop yields drop significantly due to the lower sunlight absorption. Another role that silica plays in plants is responding to biotic and abiotic stresses. For example, plants can use silica to contain infections occurring on leaves by depositing silica at the infected locations [12].

6.2.4 Bio-magnetite: sensing, cutting/grinding, iron storage

Although not so well known, iron oxides are the other main class of biominerals. There is a large variety of iron oxide minerals, and of these, only a few are biomineralised. These are goethite (α-Fe^{3+}OOH), lepidocrocite (γ-Fe^{3+}OOH) and ferrihydrite ($Fe^{3+}_{10}O_{14}(OH)_2$), which are all ferric minerals, and as such precipitate out of a solution readily. Magnetite ($Fe^{3+}_2Fe^{2+}O_4$) is the final remaining biomineralised iron oxide, which is the only mixed valence example, meaning it requires more control from the organism to form it. Iron oxides are generally very hard and dense and can be magnetic (in the case of magnetite), and thus are not deposited in organisms for structural purposes; these characteristics are more commonly utilised to function as teeth and magnetic sensors. As such, in most cases iron-oxide biominerals do not contain organic/bio molecules. Limpet teeth, used for scraping and grinding, are biomineralised goethite, while chiton teeth are formed from lepidocrocite or magnetite, with an underlying core of ferrihydrite, which is proposed to be the precursor phase to the harder mature surface layer of magnetite.

Magnetite nanoparticles and larger structures occur in a wide variety of higher organisms such as honeybees, pigeons, pelagic fish, bats, rodents and humans, with approximately 5 million crystals per gram of human brain tissue. Although the

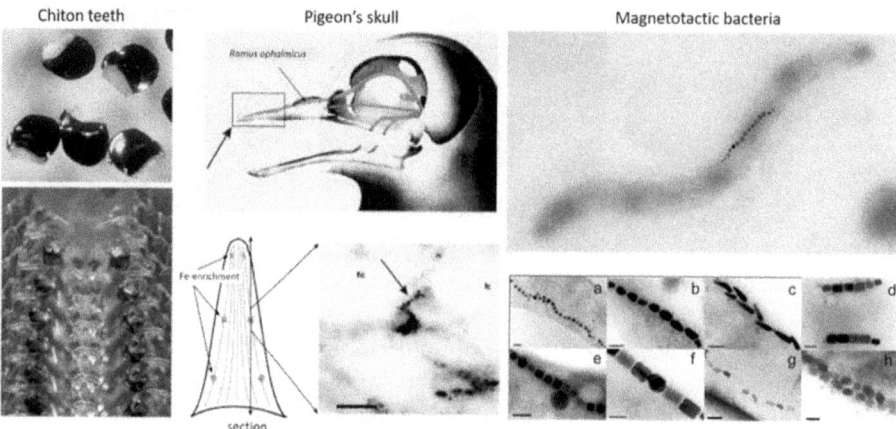

Figure 6.6. Three examples of biomineralisation of magnetite. Left-hand side: chiton teeth coating in magnetite [3] (© IOP Publishing, reproduced with permission. all rights reserved) (top); teeth along the radula showing magentite coating in the forefront (bottom of the image) and without in the immature teeth (at the top of the iamge) (bottom) (from [14], reprinted by permission from Macmillan Publishers Ltd). Centre: presence of nanoparticles of magnetite within beak of a pigeon ([15], reprinted by permission from Macmillan Publishers Ltd). Right-hand side: magnetite nanoparticles biomineralised within magnetotactic bacteria (top); comparison of the range of different morphologies of these nanoparticles (bottom) (reprinted from [16], copyright 2012 with permission from Elsevier.

purpose of these particles is not yet fully understood in all cases, the majority are proposed to be for sensing/navigational use, acting as a natural compass, helping bees and pigeons (figure 6.6(B)) find their way home, for example. However, an excess of magnetite in the human brain has been linked to neurodegenerative diseases such as Alzheimer's and Parkinson's. Magnetite is formed on the surface of chiton teeth to provide a hard material for scraping food from rocks (figure 6.6(Ai)). This is an excellent example to study, as the teeth are constantly being produced and replaced along the tongue-like radula the chiton uses for feeding, with the immature teeth at the back being developed in time to mature to take the place of worn teeth at the front. As such the teeth are at different stages of development along the radula row (figure 6.6(Aii)), so the mineral can be probed over its complete formation from just one sample [14] and shows that the magnetite is formed from a ferrihydrite precursor. The occurrence of intracellular magnetite in lower organisms is limited to a few bacteria, which are magnetic. Magnetotactic bacteria are motile and live in aqueous environments (fresh and salt water) (figure 6.6(Ci)). They take ferrous ions up from the water and use this to biomineralise highly monodispersed nano-crystals of magnetite in liposomes (biological membrane vesicles) within their cells (called magnetosomes) arranged in a linear chain. Remarkably, there is a range of magnetite shapes and sizes across different species, but the shape and size is tightly conserved within each strain, showing the high level of biological control (figure 6.6(Cii)). While these magnets align the bacteria to the Earth's magnetic field, and a navigational function is commonly accepted, there could be other reasons for biomineralising magnetosomes, such as a metabolical function, or detoxifying a harmful iron radical.

Iron oxides can also be deposited primarily for the storing of large quantities of essential iron such as ferrihydrite, which can then be redissolved when required. This is performed within a protein called ferritin, which assembles into a large symmetric hollow protein ball. This protein imports iron ions that precipitate as ferrihydrite within the core. Ferritin is very common across the full range of organisms from bacteria to humans.

6.3 Mineral formation controlling strategies in biomineralisation

The most incredible feature of biomineralisation is the biological control and regulation of the entire process, from the sequestering of ions and molecules from surroundings to the deposition of stunningly beautiful, organised and often multi-component complex structures that are described in the last section. It is this that clearly separates biominerals from chemically synthesised inorganic materials. While there is a diverse range of biominerals and an even more diverse range of properties and functions that these minerals can offer, the process of biomineralisation can be simplified to consider three essential generic steps that are universal throughout (figure 6.7). These are:

1. The sequestering of the reagents required for biomineralisation, and the concentration of these.
2. The nucleation of the specific inorganic mineral.
3. Control of the crystal growth, specifically with respect to the morphology of the final mineral.

The key to biomineralisation is how Nature uses biological components to control and regulate the chemical processes that occur in these steps.

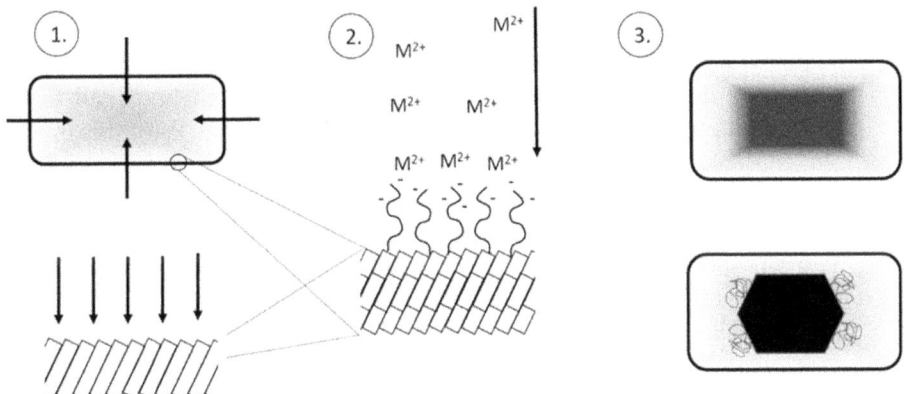

Figure 6.7. Schematic of a generic biomineralisation process. Copyright (2006) National Academy of Sciences [17].

6.3.1 The universal biomineralisation process

(1) While the minerals and thus the reagents vary, the principles of sequestering the cation are the same and basically rely on reagent concentration so precipitation can occur. Confinement is required if you want to concentrate a particular species in that solution in a defined space, compared to the bulk solution. Thus the concentrated cations need to be segregated before biomineralisation occurs, so they can be released at a specific location for mineralisation to take place.

This generally follows one of two processes depending on the size of the mineral. On the larger scale (>internal space within a cell) mineralisation occurs on an extracellular organic matrix, while smaller particles can be mineralised intracellularly within discrete vesicles (on the nanometre scale), while vesicles can also be used in extracellular biomineralisation to concentrate the mineralising species and traffic them to the mineralisation sites where they are deposited, allowing the larger extracellular mineralisation to occur. The cations required are sequestered from the environment often taken up into vesicles. Extracellular mineralisation can occur without the need for vesicle trafficking, if the concentrations of the mineral ions in the environment the mineralisation site is exposed to are high enough to simply precipitate out of the solution onto a nucleation site on the template or surface (see below (2)). In this case, larger scale compartmentalisation occurs in the form of restriction of crystal growth and thus size by features such as cell walls and/or large organic template boundaries (see below (3)).

(2) Biominerals form on biological surfaces (i.e. heterogeneous nucleation) as this has a lower formation energy barrier than in homogeneous nucleation. Hence the formation of biominerals on surfaces is favoured. Most biominerals are precipitated easily from solution and are very thermodynamically stable. This is not a coincidence. Biology will tend to utilise the most accessible chemistry. Thus in many cases simply concentrating the reagents on a surface will result in the correct biomineral being formed. In fact, many calcium minerals precipitate readily and our human biology very carefully regulates the amount of calcium that is absorbed into the blood, and requires biomineralisation *inhibitor* proteins to prevent unwanted mineralisation in the wrong places. We may be familiar with hardened arteries through unwanted calcification, leading to heart disease. With this in mind, the purpose of a biomolecular nucleation site is not simply to induce nucleation, but to induce nucleation preferentially at the correct site, to grow in the correct orientation and of the correct mineral. There are also two aspects to the mineral type: the chemical composition and the structural polymorph (e.g. within the same environment, shells produce both aragonite and calcite using different proteins to direct the formation of different polymorphs). Typically these proteins will preferentially bind the required cations such as Ca^{2+} or Fe^{3+} and this will direct the formation of a particular mineral phase and even orientation. Nucleation proteins are

thus found on the surface of the mineralising site either on the extracellular matrix or embedded in a vesicle membrane. Sometimes the mineralising surface intrinsically has this nucleating functionality so that an additional protein or other biomacromolecule is not required.

(3) There are several ways in which biology achieves various morphologies of the final biominerals. The first is the use of a template. There are two types of template. One is a mould, to grow the biomineral in a confined space of a specific morphology, like ferrihydrite inside the confined ferritin core. The second is to mineralise onto a biological structure as a template, so the mineral forms a shell of the same shape as biological matrix. A further way to control mineral morphology is to control crystal growth by inhibiting growth of specific crystal facets, leading to these facets dominating the final morphology as the fast growing planes grow out to extinctions. Growth of certain crystallographic planes can be inhibited by selective biomolecules binding. Finally, the organism can use chemical control to convert a modellable amorphous phase into a crystalline functional phase once the morphology is achieved, and hence phase transformation is performed *in vivo*. For example, spicule formation in sea urchins occurs via the initial deposition of amorphous calcium carbonate that is 'moulded' into the final required morphology, before being converted into the crystalline form—calcite.

6.4 Roles and types of organic biological components required for biomineralisation

In this section we will discuss the characteristics and properties required to perform functional roles, and the types of biological components that fulfil these in Nature.

6.4.1 Roles of organic biological components

From the three steps outlined above we can clearly identify the roles of biomolecules/organic components/matrices that are required. Broadly, these can be simplified as:
- confinement, concentrating (trafficking) of reagents;
- nucleating;
- controlling morphology: templating onto a matrix surface;
- controlling morphology: crystal growth control (through confinement or epitaxial growth).

6.4.1.1 Reagent confinement/concentrating trafficking
The role of biological components in this step is to form a barrier to divide up the physical and chemical space, to enable reagents and precursors to be isolated, concentrated and moved to other sites. Nanoscale vesicles can feature in both macro and nanoscale process. For intracellular biomineralisation, the vesicle serves as a miniature nanoscale reaction vessel: a place for the reagent to concentrate and mix. Therefore the vesicle must be formed of non-permeable membrane, such as phospholipid, or a fibrous membrane with a specific ion transporter protein

embedded, that selectively transports the mineralising reagents in (concentrating them), and the by-produce out, to ensure the correct chemistry is maintained and there is a kinetic driver for precipitation to occur.

6.4.1.2 Nucleating
Typically these are biomolecules and biomacromolecules, with highly acidic side-chains such as proteins with multiple aspartic and/or glutamic acid residues, or sugars, with multiple carboxylic acids and polar hydroxyl groups that will bind the required cations. Note that repeating acidic motifs are the most common biomolecules for the biomineralisation of calcium and iron minerals, whereas the chemistry of silica precipitation differs and thus polar residues such as cysteine, threonine and serine are frequently included, and proteins dominated by basic (cationic) residues, such as lysine and arginine, are utilised in this system to promote the precipitation of silica from silicic acid.

Multiple binding sites are required to give a charged surface to nucleate the formation of a mineral (rather than just binding discrete metal ions), so a protein/polymeric macromolecule must have multiple bind sites or smaller nucleation proteins will aggregate together, and this will direct the formation of a particular mineral phase. The location of these proteins can also orientate the crystals to grow in a specific direction or orientation. Interestingly, these proteins tend to be intrinsically disordered proteins, meaning they are unstructured and flexible, which is advantageous as they will be able to accommodate the growing crystal while still being bound.

6.4.1.3 Controlling morphology: templating onto a matrix surface
Biomineralising on a biological matrix template is used to control the shape of biominerals over a vast range of sizes, from micro (coccoliths) to metres (large mammal bones). Templating biological components are usually tough fibrous macromolecules that self-aggregate to form larger matrices. Often these are made up of multiple organic molecules in complex hierarchies. In a coccolithophore for example, an organic base plate serves to direct the formation of the initial calcite minerals into a ring at the edge of the base plate (proto-coccolith). These structural template matrices are usually multifunctional, being a structural rigid template but also having active nucleating functional groups on their surface, such as aspartic acids presenting negatively charged sites for metal ion binding, so separate nucleation proteins are not required.

6.4.1.4 Controlling morphology: crystal growth control (through confinement or epitaxial growth)
This is mainly controlled by confinement. i.e. a biological component will form a confined space for mineralisation to occur, and once filled no more regents can be added and mineral growth stops. The confining biological components are standardly vesicles on the nanoscale and cell wall or other extracellular confining compartments at sizes typically over a micron, such as aragonite mineralisation envelopes in nacre. However, epitaxial crystal growth can be controlled by small

protein and biomolecules that have a binding surface with functional groups, and spacing that complementarily matches the surface of specific mineral crystal planes and steps, to bind strongly and selectively to these surfaces, inhibiting growth. As this sort of shape control relies on inhibiting selective crystal growth, it can be difficult to identify these proteins from general biomineralisation inhibitory proteins. For example, cationic lysine residues will bind strongly to most oxide surfaces, but this is also common to proteins that help inhibit the growth of kidney stones (calcium oxalate). The smaller the morphological-controlling molecules and proteins are, the fewer complications there are with surface matching and mobility in reaching the growing crystal. However, some systems use large molecular crystal matching organic matrices to both nucleate or terminate a specific epitaxially growing crystal.

6.4.2 Types of organic biological components

Some organic components are listed below with their functions (figure 6.8).
- Liposomes: concentrating/confinement, trafficking, templating.
- Polysaccharides: confinement, templating.
- Large proteins: templating, nucleating, mechanical properties.
- Small acidic proteins and macromolecules (anionic): nucleating Ca and Fe minerals.
- Small basic proteins and macromolecules (cationic): nucleating silica, crystal growth regulating.
- Cage proteins, e.g. ferritin: concentrating/confinement, templating, nucleating.

It is clear that many type of biomolecules/components with similar properties can perform the same role, while one organic biomineralisation component can carry out one, several or all of the functions above. In this section we will consider some examples of organic components.

6.4.2.1 Lipid membranes: liposomes (figure 6.8)
Lipid membranes are the primary barrier in biology and as such have a wide variety of functions in biomineralisation. Lipids are amphiphilic molecules, which means they have a long organic hydrophobic 'tail' end, and a hydrophophilic, more polar or charged 'head' group. In a polar solvent like water, lipids self-assemble into a bi-layer membrane, where two 'sheets' of lipids assemble to sandwich the hydrophobic tails between them, shielding them from the water, leaving the hydrophilic head groups exposed. Biological membranes are mainly made up of phospholipids which means that the head group is a phosphate. In eukaryotic cells there is also a substantial amount (approximately 30% of animal cells) of sterols (e.g. cholesterol), which have a hydroxyl polar head group attached to bulky cyclic hydrocarbons leading to the tail groups which can bend the membrane. There is also a small amount (a few %) of glycolipids, which have sugar head groups (monosaccharide or oligosaccharide). Such membranes are the main constituent of cell membranes across biology, both intracellular and at the cell periphery, forming cell walls. On the

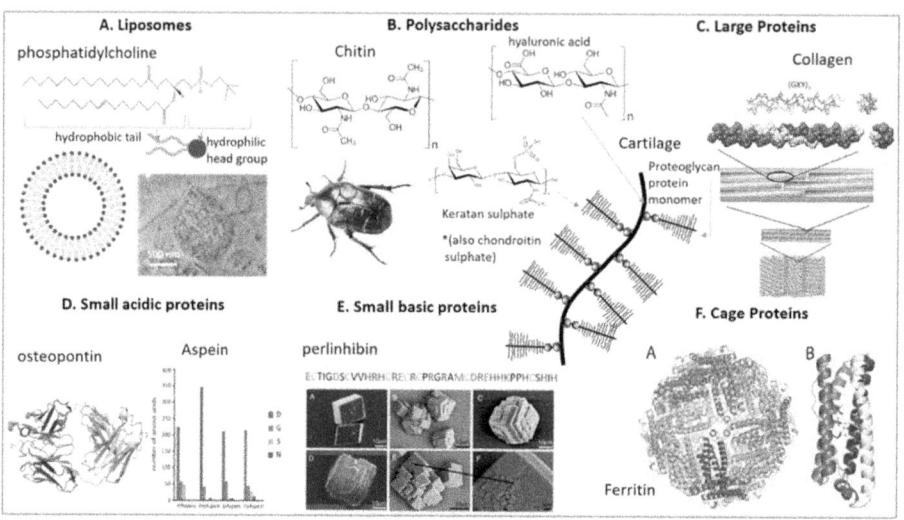

Figure 6.8. Examples of the organic components in biomineralisation. (A) Liposomes. Top: the chemical structure of a typical amphiphilic lipid molecule, phosphatidylcholine, with an uncharged hydrophobic tail on the left and depicted in orange in the cartoon representation, and the charged hydrophilic, phosphate-containing head group on the right-hand side of the molecule, depicted in blue on the cartoon representation. Below: a cartoon showing how the lipids self-assemble into a liposome in water alongside a TEM image of a calcium transporting liposomes for biomineralisation in a sea urchin larvae (reproduced from [18]). (B) Polysaccharides. Left: chitin showing the chemical structure of the monomer unit, and below this a jewel scarab beetle. Right: two further polysaccharides; the top shows the chemical structure of the monomer unit of hyaluronic acid and below the chemical structure of the monomer unit of keratan sulphate. Both of these polysaccharides are building blocks in the cartilage shown to the right. Cartilage is a biological material made up of both polysaccharides and large proteoglycan proteins. (C) Large proteins. The example of collagen is shown from the top, the triple amino acid chain structure is shown and the hierarchical structure is built up below (composed using images from [19] and [7]). (D) Small acidic proteins. Left: the protein crystal structure of osteopontin (from pdb file). Right: a graph to show the huge dominance of the acidic residues in Aspein (reproduced from [20], copyright 2012 with permission of Springer]). (E) Perlinhibin: showing the amino acid sequence highlighting the basic residues in blue, the acidic in red and the polar in green. Below: the affect perlinhibin has on calcium carbonate crystallisation and growth. (A) shows crystals in the absence of protein, with (B)–(F) in the presence of protein (reproduced from [21], copyright 2007 with permission from Elsevier). (F) Cage proteins: showing the protein crystal structure of ferritin (reproduced from [22] with permission from ASBMB).

nanoscale, lipid membranes self-assemble into spherical vesicles called liposomes. Liposomes are well utilised in biomineralisation. First they are used as compartments for sequestering, concentrating and trafficking metal ions such as calcium or iron in extracellular biomineralisation. They can also be used to traffic amorphous calcium phosphate and calcium carbonate precursors to the site of mineralisation to mature into the final mineral once deposited. This has been seen in fish bone formation and sea urchin larval spicules. They can also be used as the site of biomineralisation itself, in the case of intracellular biomineralisation. In this case the liposome serves as a miniature nanoscale reaction vessel: a place for the reagent to concentrate and mix. The liposome membrane has specific ion transporter proteins embedded, which transport the mineralising reagents in (concentrating them), and the by-produce out, to create a chemical gradient across the membrane to ensure that the correct chemistry is maintained and that there is a kinetic driver for precipitation to occur. The CAX family of membrane proteins is a well know calcium ion transporter. These are antiporters, that take Ca^{2+} into the liposomes and export H^+ out. CAX3 is critical in the uptake of Ca^{2+} into liposomes in coccolith scale biomineralisation. MamM and MamB are other liposome membrane protein antiporter systems for the uptake of iron ions into the magnetosome (and the export of H^+). Biomineralising vesicles (such as magnetosomes (specifically discussed further in chapter 8)) also contain a suite of proteins with nucleating site (see below) redox proteins to maintain the chemistry required for mineral formation and shape controlling proteins. Collectively, liposomes can also form an external boundary template, with the lace-like structures of silica forming between liposomes in diatoms.

6.4.2.2 Polysaccharides (figure 6.8)
A saccharide or monosaccharide is the single heterocyclic hydrocarbon unit of sugars. They are a ring composed of carbon with one oxygen, with multiple hydroxyl groups. Polysaccharides or oligosaccharides are composed of chains of these monosaccharide units, polymerised to form a range of sugars and carbohydrates. These large polymer macromolecules are well utilised in biomineralisation due to the way in which: (1) they can assemble into strong yet soft structures and matrices (such as chitin) so can have a templating function, and (2) the hydroxyl and deprotonated acid hydroxyl can coordinate to the mineralising metal ions (such as Ca^{2+} and Fe^{3+}) to offer repeating nucleation sites. Furthermore, these fibrous structural components conjugate to other organic components, readily displaying liposomes and smaller nucleation proteins, while also contributing to many more complex hierarchical organic systems (see section 6.4.2.3). Chitin is commonly used in biomineralisation as a template. It forms the flexible matrix within which to form the iron oxide—goethite needle crystals. This exceptionally tough hybrid organic/inorganic material is used to form limpet teeth. Similarly, the organic matrix of the cuttlefish bone and the templating structure of nacre in sea shells is comprised of chitin. A critical biomineralisation platform is the base plate where the coccolith is assembled in coccolithophores. This is composed of mainly the polysaccharide cellulose, where several proteins attached are responsible for binding calcium ions.

6.4.2.3 Large assembly fibrous proteins (figure 6.8)
Most large assembly proteins have a mainly structural function in biomineralisation, forming a template or scaffold for the mineral to form on or within. However, they also usually have functional amino acids that aid mineralisation upon them. One example is the glycoproteins called frustulins, which aid the formation of the silica diatom shell. These have large areas of hydrophobic amino acid residues (repeats of glycine with numerous proline and tryptophan residues) making them form insoluble specific structures (see collagen, below) identifying the proteins to have a

dominant structural framework function. However, an acidic/polar region is also present that identifies as interacting with the forming mineral, with a repeating motif of [Cys-Glu-Gly-Asp-Cys-Asp]. A mix of polar and acidic residues are important for the interaction and mediation of silica biomineralisation. Another class of proteins (HEP) closely associated with frustulins also shows a high quality of polar and acidic residues, with serine and threonine making up a quarter of its sequence, and a fifth of the sequence being glutamic and aspartic acid.

There are also several large proteins that assemble into insoluble fibrils for extracellular templating for biomineralisation. The most common structural protein of this type is collagen. Collagen is a large (>1000 amino acids) protein made up of a repeating sequence of glycine, proline and other amino acids, to such an extent that every third amino acid is glycine, with proline making up a sixth of the total residue composition. This makes the protein hydrophobic and the combination of proline and glycine bends the protein backbone giving the protein a regular twisted structure that assembles into trimers of three twisted strands: tropocollagen, akin to a rope 280 nm long and 1.5 nm wide. Interestingly, collagen contains a large quantity of proline with a hydroxylation post translational modification (9% of the amino acids in collagen are hydroxyproline) and also some lysine. It is the addition of these hydroxyl groups that helps these protein rope units to assemble yet further by cross-linking them in a staggered fashion into long insoluble collagen fibrils. This hierarchical assembly gives the fibrils a building-block unit structure, with gaps between the ends of each trimer twisted strand and structural grooves between them. These have huge importance as the defined pockets (known as the 'hole zones') of nucleation, making up regular nucleation sites along the fibril. Specifically, hydroxyapatite crystallises only within the hole zones initially and the crystals grow along the grooves, so the crystal growth is preferentially orientated. This is a clear example of how organic hierarchical structures can template materials with both short- and long-range structural order.

An example that combines both larger proteins (proteoglycan) and polysaccharides (hyaluronic acid, glycosaminoglycan) to form a complex organic material that interacts and interweaves with collagen is cartilage. Cartilage is the main component of bone when we are born and as such forms the macro-level template of the bone shape. It is tough, smooth and bouncy, so ideal for light youngsters prone to accidents and falling, who don't yet require fully mineralised bones. Collagen and thus mineralised bones form on the cartilage templates over time, so once the bone matures, it entombs the cartilage inside, which then degrades away to leave a hollow void that more bone structures can form in. Cartilage does remain to cover the surface of the bone to protect it against impact, particularly at the joints. The structure of cartilage is very complex and resembles the hierarchical structure of a fern frond: it has a long chain polysaccharide (hyaluronic acid) core (like the central stem), from which numerous proteoglycans extend out perpendicularly at regular 30 nm intervals (as the pinna would). More polysaccharides (around 80) are perpendicularly linked along the length of this protein (resembling pinnules). The repeating units of these carry a negative charge resulting in a large amount of hydration (H-bonding to water) in cartilage, as well as good anionic sites for interaction with collagen and for Ca^{2+} binding to nucleate mineralisation.

6.4.2.4 Small acidic proteins and macromolecules (anionic) (figure 6.8)
There is a vast range of acidic mineralisation proteins, and most of these are nucleation proteins that bind metal ions into a surface to drive spontaneous heterogeneous nucleation. Examples include osteopontin and osteonectin, which are both nucleation proteins in bone that are very acidic residue rich, with large amounts of glutamic and aspartic acid in osteonectin, while osteopontin has a 9 aspartic acid repeat sequence. Aspartic acid repeat sequences are particularly prevalent in biomineralisation nucleation proteins, in a range of systems from bone to a full range of sea shells such as barnacles and mussels. For example, the C-terminal of the calcite nucleation protein aspein is almost entirely aspartic acids. Combined glutamic and aspartic acid rich motifs and repeating sequences are commonplace, and where the sequence is not repeated, nucleation proteins contain motifs to enable aggregation so the acidic motifs are repeated across the aggregate surface. One example is Mms6, from the magnetosome in magnetic bacteria, which nucleates magnetite by binding both ferrous and ferric ions on an Asp-Glu-Glu-Val-Glu motif. This also has a Leu-Gly repeating hydrophobic motif, which promotes the aggregation of the protein through a bulk–non-bulk group—a knob and hole type interlocking interaction and this is further described in chapter 8. Another method of obtaining repetition for a small acidic nucleation protein is to assemble at specific sites on a regular organic surface, such as another protein (like the regular β-sheet structured silk-fibroin-like protein in nacre biomineralisation, or the polysaccharide chitin which occurs in limpet teeth formation (see section 6.4.2.2)).

Further to the nucleation function, small acidic proteins can also be used to control the crystal habit and epitaxial growth of biomineral crystals. This relies on the charge on the surface mapping onto the complementary charged surface of the mineral. Acidic biomineralisation proteins from adult sea urchins interact specifically with the prismatic face of Mg-calcite, while small acidic abalone nacre protein (AP8) directly control calcite morphology to be asymmetrically curved when added to chemical precipitation of this mineral *in vitro*.

6.4.2.5 Small basic proteins and macromolecules (cationic) (figure 6.8)
While it is more uncommon to find cationic proteins or protein motifs in biomineralisation (compared to acidic proteins and motifs), there are examples of basic amino acids controlling the formation of particular crystal phases or morphologies, interacting more strongly with the forming mineral face rather than the precursor ions. Two examples are the cationic/polar region at: (1) the C-terminus of the Japanese pearl oyster shell matrix protein, prismalin 14, which is found at the inner side of the outer face of the shell mantle, and (2) the N-terminus of the nacre associated sea snail shell protein, perlinhibin. Both proteins bind very strongly to the forming mineral, regulating the crystal formation of the prismatic layer of the shell. Both proteins also 'inhibit' mineralisation *in vitro* which may demonstrate a preferential crystallisation function. A third example of basic proteins

are the highly cationic proteins found in diatoms containing arginine and lysine residues. These proteins also contain propylamine modifications on various residues, making them even more basic. They strongly interact during the nucleation, particle formation, growth and assembly stages in biosilica formation—this is further detailed in chapter 9.

Silaffins are a group of small basic proteins found in silica biomineralisation of diatoms. Again these are enriched for the cationic residues lysine and arginine, along with hydroxyl polar residues serine and threonine. These functional groups tend to have regular repeating motifs and as such the proteins interact with silicic acid and bind strongly to the forming silica, suggesting a crystallisation role.

6.4.2.6 Cage proteins (figure 6.8)

The cage protein ferritin is a remarkable example of a single protein that performs all the functions of biomineralisation for the purpose of iron storage and slow release. Its multi-functionality stretches from sequestering iron ions into its confinement/templating core, nucleating the mineral ferrihydrite and regulating the size of this mineral to the interior core of the protein. Ferritin is a ubiquitous cage protein found in almost all living organisms. Iron-mineral-free mammalian ferritin (apoferritin) consists of 24 subunits, which are each made up of four helix bundle assembles. These subunits assemble to form a dodecameric cage 12–14 nm in diameter. The symmetric structure of this cage gives six pores into the hollow core which act as ion channels, importing the soluble ferrous iron ions. Within the interior of the protein the ferrous ion is oxidised to ferric iron ions at a ferroxidase site on some of the helices. The now higher charge ferric ions are then electrostatically attracted to a highly negatively charged binding pocket, accessed on the cage interior between the four helices of each subunit bundle. This pocket nucleation site is lined with five glutamic acids, a histidine and glutamine to bind two iron ions and a water molecule. The reagents then cross-link to form the ferrihydrite mineral. This grows from each of the nucleation sites until the ferritin is filled with ferrihydrite. The cage protein ferritin is an excellent example of all biomineralisation components in one protein. It encompasses compartmentalisation, behaving as nanoreactor which can sequester the iron ions and then nucleate and direct the mineral growth. Such cage proteins and their mimics show themselves to be very useful in bioinspired green synthesis of nanomaterials seen in chapter 7.

6.5 Summary: key lessons from biomineralisation for the green synthesis of nanomaterials

It is clear that biomineralisation can occur over the whole range of length scales. While it may seem that micro- and macro-level examples of biomineralisation may not be relevant to the formation of nanoscale materials, this is simply not the case, as common trends and features across the whole range of sizes can be applied to bioinspired nanomaterial synthesis. Furthermore, macro-biominerals are hierarchical across the length scales, so have nanoscale intricacies and precision. The

common themes that can aid the design of bioinspired approaches to making nanomaterials are the ways in which biology controls:
1. The chemistry of the environment on the nanoscale which affects the nanomaterials formed. This is controlled by ion pumps and redox proteins in biomineralisation, but could utilise other approaches synthetically.
2. The confinement of crystal growth which directs the shape and size of the materials produced, such as formation within liposomes.
3. Organic molecules are adept at forming into a full variety of shapes and architectures at all length scales. However, these are always fundamentally controlled at the molecular level, for example by protein sequence features which introduce bulky amino acid residues that cause a bend in the protein shape. These form intricate scaffolds to template the formation of very specific shaped materials.
4. Patterned arrays of positively charged functional groups can nucleate a specific material by binding metal ions to such an extent they can control the formation of a specific crystal phase and can even direct growth through nucleation of a specific crystal face.
5. Small soluble charged proteins and biomolecules that bind to the forming mineral at specific steps and faces inhibit the growth of these sites, and thus control the morphology of the resulting crystal at the nanoscale.
6. Mixtures of organic and inorganic (hybrid) hierarchical materials can show superior physical properties, which could be utilised when designing new nanomaterials.

References

[1] Behrens P and Baeuerlein E 2007 *Handbook of Biomineralization: Biological Aspects and Structure Formation* (Weinheim: Wiley), p 412
Mann S 2001 *Biomineralization: Principles and Concepts in Bioinorganic Materials Chemistry* (Oxford: Oxford University Press), p 240
[2] Weiner S and Dove P M 2003 *Rev. Mineral. Geochem.* **54** 1
[3] Addadi L and Weiner S 2014 Biomineralization: mineral formation by organisms *Phys. Scr.* **89** 098003
[4] Fox D 2016 What sparked the Cambrian explosion? *Nature* **530** 268–70
Tarhan L G, Droser M L, Cole D B and Gehling J G 2018 *Integr. Comp. Biol.* **58** 688
[5] Maloof A C, Porter S M, Moore J L, Dudás F Ö, Bowring S A, Higgins J A, Fike D A and Eddy M P 2010 *GSA Bull.* **122** 1731
[6] Kocot K M, Aguilera F, McDougall C, Jackson D J and Degnan B M 2016 *Front. Zool.* **13** 23
[7] Zimmermann E *et al* 2016 Intrinsic mechanical behavior of femoral cortical bone in young, osteoporotic and bisphosphonate-treated individuals in low- and high energy fracture conditions *Sci. Rep.* **6** 21072
[8] Von Dassow P, Díaz F, Bendif E M, Gaitán-Espitia J-D, Mella-Flores D, Rokitta S, John U and Torres R 2017 Over-calcified forms of the coccolithophore Emiliania huxleyi in high-CO_2 waters are not preadapted to ocean acidification *Biogeosciences* **15** 1515–34
[9] Aizenberg J and Hendler G 2004 *J. Mater. Chem.* **14** 2066

[10] Taylor A R, Brownlee C and Wheeler G 2017 *Annu. Rev. Marine Sci.* **9** 283
[11] Round F E, Crawford R M and Mann D G 1990 *The Diatoms: Biology & Morphology of the Genera* (Cambridge: Cambridge University Press), p 747
[12] Kauss H, Seehaus K, Franke R, Gilbert S, Dietrich R A and Kroger N 2003 *Plant J.* **33** 87
[13] Patwardhan S V, Clarson S J and Perry C C 2005 *Chem. Commun.* **9** 1113
[14] Gordon L M and Joester D 2011 *Nature* **469** 194
[15] Hanzlik M, Heunemann C, Holtkamp-Rötzler E, Winklhofer M, Petersen N and Fleissner G 2000 Superparamagnetic magnetite in the upper beak tissue of homing pigeons *Biometals* **13** 325–31
[16] Staniland S S, Tolosa J, Wilson O, Garcia-Martinez J C, Binns C, Rawlings A E and Bramble J P 2014 *Nanomagnetism: Fundamentals and Applications* ed C Binns (Amsterdam: Elsevier)
[17] Buehler M J *et al* 2006 Nature designs tough collagen: Explaining the nanostructure of collagen fibrils *PNAS* **103** 12285–90
[18] Vidavsky N, Addadi S, Schertel A, Ben-Ezra D, Shpigel M, Addadi L and Weiner S 2016 *Proc. Natl. Acad. Sci.* **113** 12637
[19] Bella J 2016 *Biochem. J.* **473** 1001
[20] Isowa Y, Sarashina I, Setiamarga D H E and Endo K 2012 *J. Mol. Evol.* **75** 11
[21] Mann K, Siedler F, Treccani L, Heinemann F and Fritz M 2007 *Biophys. J.* **93** 1246
[22] Pfaffen S, Abdulqadir R, Le Brun N E and Murphy M E P 2013 *J. Biol. Chem.* **288** 14917

IOP Publishing

Green Nanomaterials
From bioinspired synthesis to sustainable manufacturing of inorganic nanomaterials
Siddharth V Patwardhan and Sarah S Staniland

Chapter 7

Bioinspired 'green' synthesis of nanomaterials

7.1 From biological to bioinspired synthesis

From chapter 6 we have learnt that one of the most interesting facets of biological nanomaterials formation is the involvement of organic biomolecules (proteins, polysaccharides, small molecules) in controlling every stage of biominerals formation, including precursor uptake, precursor/cluster solubility, transport of precursors and mineral formation. It is worth noting the mild conditions used (e.g. ambient temperatures and the use of eco-friendly reagents) [1]. A tangible example we have seen for controlled 'green' synthesis is the formation of bones. This is in striking contrast to the synthetic processes currently used both in industry and in research labs, as discussed in chapter 4. The differences between biological and abiotic routes are on a range of points, as listed below:

- The precursors used in biomineralisation are dissolved ions at low and non-toxic concentrations, versus the hazardous chemical derivatives used in lab/industrial processes.
- Water is the natural solvent for biomineralisation, while a large number of synthetic methods for nanomaterials use non-aqueous solvents or co-solvents.
- Another key difference is the temperature—biominerals are formed at physiological temperatures while abiotic methods invariably need higher temperatures.

A simple example of porous silica can illustrate these differences. As we have seen in chapter 6, diatoms deposit silica that is porous at multiple length scales, due to the hierarchical structure of the constituent particles. Yet, this deposition occurs at ambient temperatures and pressures, in water, and without needing any hazardous chemicals (it simply uses dissolved rocks as precursors). However, in order to make synthetic porous silicas to match biosilica features (e.g. to make mesoporous silica or silica with multiple porposities), hydrothermal conditions are often required (e.g. elevated temperatures between 60 °C and 150 °C and higher pressures), over long durations (hours to days). Those syntheses also commonly utilise hazardous precursors such as alkoxysilanes and organic co-solvents such as alcohols, dimethylformamide and tetrahydrofuran.

So, the question is, how can we translate the learning from biomineralisation, in order to design biologically inspired greener/sustainable routes to nanomaterials?

We note here that the purpose of designing bioinspired strategies is to create sustainable production technologies for desired nanomaterials. It is not necessarily the aim to mimic the biominerals in their structure or form, rather it is to mimic the process/principles. The main reasons for focusing on learning the principles instead of copying the structures of biominerals are twofold. First, with some exceptions, biominerals generally do not have a direct use in commercial applications, or cannot be harvested at the scale required for commercial use. It is thus important to learn the principles and build the methods to produce large quantities of nanomaterials which are desired for applications. Second, learning the principles is a powerful strategy, because it is transferrable to a diverse range of materials/chemistries, even to those that are not found in biominerals. An example of this is titania, a technologically important material in pigments and photocatalysis, which is not biomineralised. However, learning from biosilica formation has led to discovering bioinspired routes to titania synthesis [2]. Hence the use of the term *bioinspired* is appropriate over *biomimetic*. For a more detailed discussion of the terminology of biomimetic, versus biokleptic, versus bioinspired, see [3].

In figure 7.1 we can see an overview of a strategic pipeline that has been successfully used to translate the learning from biomineralisation to develop bioinspired methods. We note that this strategy could be used in an iterative approach in order to refine the learning obtained. This chapter is built and structured around this scheme as a central concept.

In this pipeline, **point 1: Biominerals and Biomineralisation** is covered in chapter 6. We have learnt about *what* biominerals are and their functions. This also includes some aspects of **point 2: Mechanistic Understanding** that biomineralisation is under strict control, where the mechanisms underpinning and the key processes involved

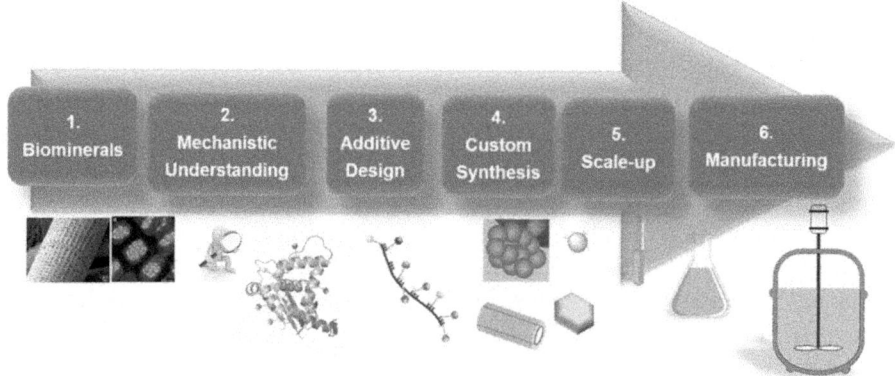

Figure 7.1. Schematic representation of a pathway from biomineralisation to bioinspired synthesis of nanomaterials.

are biologically regulated. The biological machinery exhibits physical, chemical and temporal control over the entire process. The physical and chemical control is relatively easy to mimic *in vitro*, while the temporal control can require highly sophisticated automated systems. Biochemical and biological research has revealed that biomineral formation involves organic biomolecules such as proteins and peptides in crucial stages (see chapter 6).

7.2 Mechanistic understanding

This mechanistic understanding (**point 2** in figure 7.1), once obtained, can be used to develop strategies for bioinspired synthesis. Hence developing an understanding of *how* biology produces high quality nanomaterials is crucial in moving forward. As we have seen, one of the key controlling features of biomineralisation is the use of biomolecules, so it is important to understand the roles that such biomolecules play in the entire process. As such, we will consider two main strategies for developing this 'nano–bio' interaction: (1) investigations of biomolecules that are directly involved in biomineralisation, and (2) investigations of *other* biomolecules that are not associated with biomineralisation but show strong interactions with (bio) minerals (e.g. peptides showing affinity towards synthetic minerals). Below, we describe the overview of both strategies and summarise the common features and learning. In section 7.3, we will elaborate these principles with a detailed example. This learning will be further enhanced with in-depth case studies in chapters 8 and 9.

7.2.1 Biomineralising biomolecules

From the numerous examples given in chapter 6, we can see that there has been an incredible amount of research undertaken to identify and characterise the biomolecules responsible for biological mineral formation. These include isolation via purification of specific proteins or enzymes and understanding their structure and function. Although the characterisation of these biomolecules has provided a significant amount of information, ideally, it is highly valuable to see the molecules in action in order to fully understand their function. Given the complexities of biological systems and the commensurate advances needed in the analytical sciences, it is extremely difficult to watch *live* biomineral formation with a resolution at molecular scales. This either slows the process of designing bioinspired synthesis or creates large barriers.

In order to address these challenges, scientists can (and have) focused on research using genetic engineering or model systems *in vitro*. Genetic engineering strategies have been able to understand the biomineral formation pathways, and identify key genes controlling biomineralisation and the roles of a range of biomolecules *in vivo*. Generally this has been done by making knock-out mutants of a biomineralising organism which results in the organism not making a specific protein and to see if missing that protein affects biomineralisations. Despite this progress, the translation of knowledge to develop bioinspired technologies that are suitable for commercial uptake has been difficult, because these studies do not provide a way to design synthetic/industrial processes. Direct production of biominerals may not be useful, as biominerals are not always the desired products for commercial applications. Secondly, using biomineralisation at commercial scales (tonnes) could be uneconomical. Further, purification of biominerals from cells/organisms is very challenging and costly.

On the other hand, research using model systems has been fruitful in the identification of mechanisms that can be utilised in developing novel manufacturing technologies. Such studies include, for example, probing the interactions between isolated and purified biomolecules and mineral precursors/mineral formation in the lab. As will be further elaborated in chapters 8 and 9 for the specific examples of Si and Fe minerals, such a strategy offers a much improved ability to monitor the mineral formation process via suitable sampling or *in situ* analytical instrumentation. It further helps understand the assembly and organisation of organic as well as inorganic building blocks during mineral formation. In all these cases, it is important to note the key length scales that are common to both nanomaterials and biomolecules, where interesting 'nano–bio' action takes place (see figure 2.1). It is this length scale that includes individual proteins and enzymes (~1–10s nm) and their self-assembled structures (10s–100s nm), as well as precursor clusters (~1–5 nm) that produce nanoparticles (10s–100s nm). Further, it is clear that the molecular interactions between nanomaterials and biomolecules are critically important.

Focusing on *in vitro* investigations and the key length scales, numerous researchers have identified key motifs responsible for binding, catalysis, and facilitating biomineral formation.

Probing these interesting features for a range of biominerals has helped develop deeper knowledge of the mechanisms that underpin the roles of biomolecules. Some of the commonly found mechanisms are given below; it is noted that in many cases, a combination of these mechanisms are also regularly observed. Some biomolecules bind the precursors (figure 7.2(a)), in order to change local concentrations, reactivities and special arrangements, which lead to a specific reaction pathway or the formation of *pre-determined* structures of nanomaterial product. An example where this is documented is of the soluble silicon 'pools' in diatoms. It has been reported that supersaturated levels of silicon are stabilised by biomolecules and

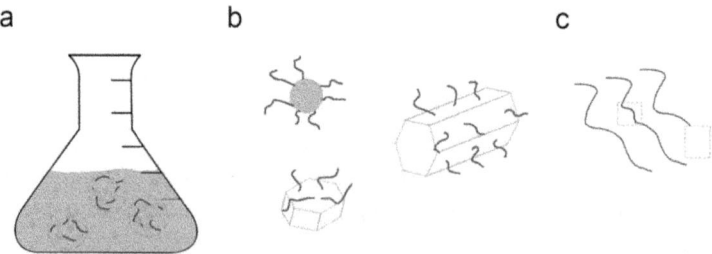

Figure 7.2. (a, b) Schematic representation of commonly observed roles of biomolecules in biomineralisation. (a) Biomolecules creating concentrated pools of precursor, and (b) biomolecules acting as capping agents. (c) One example of using site-specific mutation of a peptide sequence, either by replacing or removing specific parts of the peptide.

transported to the mineral formation sites within the organism. Other mechanisms include where the biomolecules are acting as reducing agents (in the case of metal nanoparticle formation), or capping agents (which stabilise a particular final crystal structure or particle size), see figure 7.2(b). Magnetite formation in magnetotactic bacteria is an important example of this mechanism, where the shapes and sizes of the magnetic nanoparticles are controlled by liposome surface-bound proteins (see chapter 8 for more details).

More specific details of the mechanisms can be further investigated by using sophisticated tools, such as site-specific mutations of the biomolecules, or the use of truncated biomolecules (figure 7.2(c)). The former involves *silencing* (or removing) selected sections of the biomolecules, or replacing them with other motifs, and then monitoring the effects on mineral formation. The latter focuses on a specific *functional* part of the biomolecules, which can be synthetically or biologically produced and tested for its performance in mineral formation. An example of this approach includes the use of a shorter peptide that contains the *functional* part, to develop understanding of how a matrix protein lustrin controls the calcium carbonate formation in shell and pearl nacre [4].

A correlation between the chemical structure of these biomolecules, their function and the properties of the thus formed materials can be created using the strategies explained above. Although the actual '**rules**' vary from system to system, studies have revealed interesting features that are commonly found in biomineralising systems:
- unique catalytic or binding *sites* in enzymes or proteins,
- specific (and sometimes repeating) *sequences* of key amino acids,
- particular *self-assembly* of these biomolecules (intra- and inter-molecular) and
- peculiar *chemical structures*, modifications or motifs.

7.2.2 Abiotic peptides and proteins from biopanning

Biopanning (explained in section 5.5.2.3) has been highly successful for a range of materials in identifying peptide sequences that show specific recognition for material surfaces. Importantly, this methodology can be used to investigate interactions of peptide and proteins with abiotic minerals just as readily as those biomineralised, allowing us to move away from biomineralisation towards bioinspired for any inorganic nanomaterial. The materials investigated using this approach range from metals and alloys (e.g. Ag, Au, Pd, Gd, Ti, Co/Pt and Fe/Pt), to oxides (e.g. of Si, Fe, Ti, Zn, Sn, Ge, Mn, Cr, Co, Pb), zeolites, sulfides, selenides, arsenides, salts, $CaCO_3$, and even organic materials such as carbons (e.g. fullerenes, graphene and nanotubes) and polymers [5]. A range of peptide sequences identified for selected materials are listed in table 7.1, along with exemplar applications emerging from exploiting these nano–bio interactions.

The key features of peptide binding on surfaces and the benefits of using these peptides, as classified into two broad areas, are given below [6]:
- Selectivity and high affinity binding.
- Ability to mediate synthesis, assembly and/or functionalisation.

Table 7.1. Sequences of nanomaterial-binding peptides for selected systems and their exemplar applications. Table adapted from [5].

Target nanomaterial	Peptide sequence	Applications
Metals		
Gold	MHGKTQATSGTIQS	Biocompatibility and synthesis/assembly
	WALRRSIRRQSY	
	WAGAKRLVLRRE	
	LKAHLPPSRLPS	
	VSGSSPDS	
Palladium	TSNAVHPTLRHL	Nanomaterial synthesis and controlled catalysis
Platinum	PTSTGQA	Nanostructure synthesis and biocompatibility
	TLTTLTN	
	SSFPQPN	
	CSQSVTSTKSC	
Cobalt platinum	KTHEIHSPLLHK [7]	Inserted into modified ferritin cages
Iron platinum	HNKHLPSTQPLA [8]	For $L1_0$ phase
	SVSVGMKPSPRP [8]	
	VISNHRESSRPL [8]	
	KSLSRHDHIHHH [8]	
Silver	AYSSGAPPMPPF	Biomaterial synthesis
	NPSSLFRYLPSD	
Metal oxides		
Iron oxide	QKFVPKSTN [9]	Cubic shape controlling
	IKKKKYKY [9]	Biomolecule immobilisation
	RRTVKHHVN	
Lanthanide oxide	ACTARSPWICG	Biocompatibility
Silica	MSPHPHPRHHHT	Nanostructure synthesis
	KSLSRHDHIHHH [10]	
	SSKKSGSYSGSKGSKRRIL	

7-6

Material	Peptide sequence	Application
Titania	HPPMNASHPHMH RKLPDA	Biocompatibility and nanostructure synthesis
	RPRENRGRERGL RKLPDA	
Quartz	PPPWLPYMPPWS	Biocompatibility
Zinc oxide	EAHVMHKVAPRP RPHRK	Controlling crystal types and morphologies [11]
Minerals		
Calcium phosphate	KDVVVGVPGGQD	Biomolecule immobilisation; biomaterial synthesis
Hydroxyapatite	NPYHPTIPQSVH	Biomaterial synthesis
Zeolites	VKTQATSREEPPRLPSKHRPG	Biomolecule immobilisation
Carbon materials		
Graphene [12]	EPLQLKM, QQQLSTH TMGFTAPRFPHY YHRMPQALSAME GAMHLPWHMGTL	Biomolecule immobilisation
Single-walled carbon nanotubes	HSSYWYAFNNKT DYFSSPYYEQLF DSPHTELP	Biocompatibility and multi-material fabrication
Semiconductors		
Cadmium sulphide	CTYSRKHKC	Multi-material fabrication and nanomaterial synthesis
Gallium arsenide	AQNPSDNNTHTH	SBP binding studies
Zinc sulphide	CGPAGDSSGVDSRSVGPC CNNPMHQNC LRRSSEAHNSIV	Biomolecule immobilisation; nanomaterial synthesis

These points are also illustrated in section 7.3 with selected examples from the literature. The selectivity observed with peptides is remarkable, and is often viewed as the ability of a peptide to *recognise* materials. A number of peptides have been identified (as shown in table 7.1) that recognise surface features of materials, and are able to distinguish small changes in the surface chemistry of the same material (e.g. ionisation) and/or crystal faces/planes. This remarkable property has enabled the modulation of solubilities and stabilities of precursors, intermediates and/or final nano-structures (see examples below in section 7.3).

This recognition ability further helps in controlling nanomaterials synthesis, assembly and even functionalisation. This can be achieved by reducing surface energies upon selective binding of peptides, and thus providing thermodynamic driving force for nucleation and growth of this one preferred phase or face (see section 4.4.3). An example of this is shown in figure 7.3 (top) in the case of ZnO crystals, where peptide binding can modulate the crystal size and shapes. As a result,

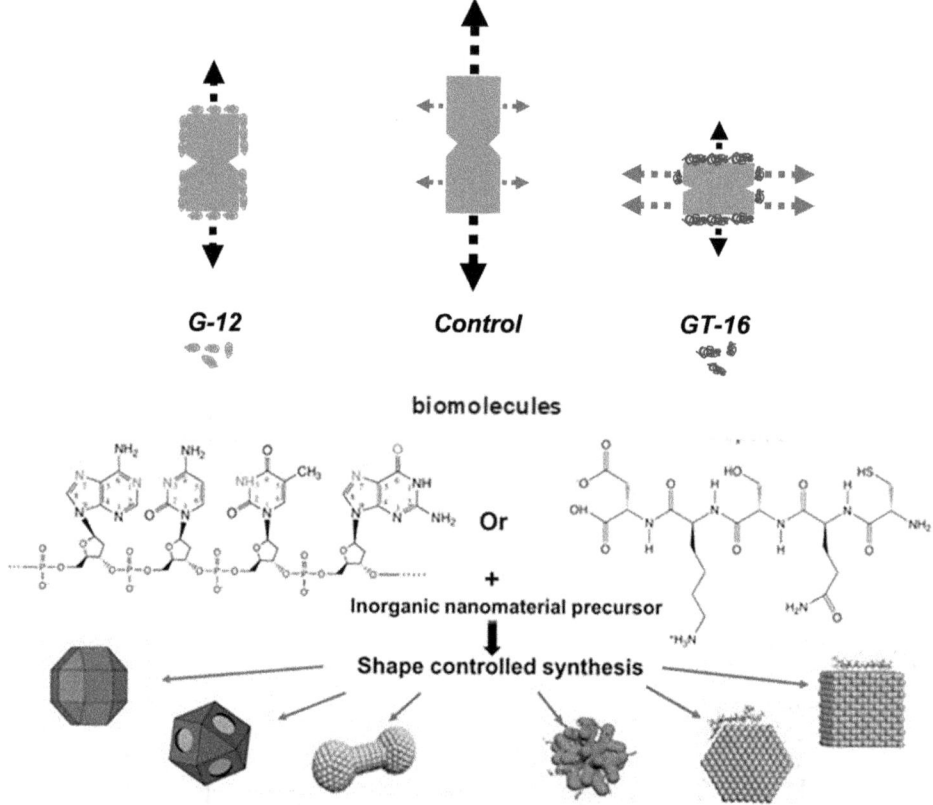

Figure 7.3. Top: schematic representation of two peptides (G-12 and GT-16) that preferentially bind different surfaces of a crystal, thus controlling growth of those faces, resulting in different morphologies. Image adapted from [11], by permission of the Royal Society of Chemistry. Bottom: the toolkit of biomolecules can be used to produce a range of nanomaterials. Image taken from [13], with permission from Elsevier.

peptides not only *catalyse* the formation of nanomaterials by reducing the energy barriers, but they can also select the particle formation pathways. In some cases, these interactions have resulted in reaching new structures that were previously not possible to achieve (see figure 7.3 and examples below).

This approach *only* provides the information on the sequences of tight binding peptides, and on its own does not shed light on the binding mechanisms, e.g. binding selectivity, affinity, rates and energies, material surface properties and peptide structure and conformation, which require further analyses. Hence additional research is required for developing the understanding of the physicochemical properties of peptides and materials using a range of experimental approaches coupled with bioinformatics and molecular modelling and simulations. A number of such experimental and computational tools have been employed successfully to probe nano–bio interactions and they are summarised in table 7.2, while extensive information on a range of techniques used can be found in [14].

7.3 An illustration of exploiting the knowledge of nano–bio interactions

In order to illustrate the key points discussed above, we will use an example of Pd nanoparticles from the literature (extensively reviewed in [15]). The aim is to follow the journey from the identification of strong binding peptides, developing the fundamental nano–bio understanding, all the way to their use in applications. In the case of Pd nanoparticles, Knecht and co-workers initially identified strong binding peptides against commercially available Pd nanoparticles [16]. The most promising peptide was Pd4 (see table 7.3 for the sequence). Using various analyses of the peptide sequence, previous knowledge of metal-binding ligands and, crucially, computational analysis, they identified histidine at positions 6 and 11 as the key residues for Pd binding. Indeed, nitrogen-containing ligands are known to bind metals, in particular those with imidazole rings as found in histidine. In order to confirm this hypothesis, they created a range of *mutant* peptides using the strategy shown in figure 7.2c, where one or both histidine residues were replaced by alanine or cysteine residues (as shown in table 7.3). Detailed experimental investigations of the peptide binding, Pd nanoparticle synthesis using these peptides, and the analysis of peptide structure and confirmation were performed.

The Pd nanoparticles synthesised using these peptides were further subjected to in-depth structural analysis at the atomic scale using x-ray diffraction, x-ray absorption spectroscopy and molecular dynamics simulations. The interpretation of results using reverse Monte Carlo analysis provided atomically resolved mechanistic details of the nano–bio interactions. Importantly, it was revealed that the Pd nanoparticles synthesised using these peptides contained an ordered core and a disordered surface (figure 7.4). Interestingly, the disordering of Pd atoms on the surface depended on the peptide (sequence and structure). These predicted structures of Pd nanoparticles strongly correlated with those observed using a high resolution TEM.

Table 7.2. Experimental and computational tools employed to probe *nano-bio* interactions. Table adapted from [6].

Tools	Affinity	Selectivity	Kinetics and thermodynamics	Surface coverage	Peptide structure and conformation	Surface properties
Fluorescence microscopy	✔	✔				
Site-directed mutagenesis	✔	✔			✔	
Surface plasmon resonance	✔	✔	✔	✔		
Quartz crystal microbalance	✔	✔	✔	✔		
Isothermal titration calorimetry	✔	✔	✔	✔		
Atomic force microscopy	✔	✔		✔		✔
Circular dichroism spectroscopy					✔	
Nuclear magnetic resonance					✔	
Fourier-transform infrared spectroscopy	✔				✔	
Computer models and simulations	✔	✔	✔		✔	✔
Zeta potential	✔		✔			✔
Raman spectroscopy	✔					✔
X-ray photoelectron spectroscopy				✔		✔

This information was used to construct systems containing Pd nanoparticles and strong binding peptides for performing molecular dynamics simulations. The results from these MD simulations were in turn used to predict the reactivity of these Pd nanoparticles in a Stille coupling reaction. In particular, the simulations computed the ease of abstraction of a surface Pd atom; such abstraction is believed to be a key

Table 7.3. Sequences of strong binders for Pd and mutant peptides. Table adapted from [16, 17].

Peptide	Sequence
Pd4	1 6 12 TSNAVHPTLRHL
A6	TSNAVAPTLRHL
A11	TSNAVHPTLRAL
A6,11	TSNAVAPTLRAL
C6	TSNAVCPTLRHL
C11	TSNAVHPTLRCL
C6,11	TSNAVCPTLRCL
C6,A11	TSNAVCPTLRAL
A6,C11	TSNAVAPTLRCL

step in Pd-catalysed carbon–carbon coupling reactions. Remarkably, the predicted reactivities (or ease of Pd abstraction) matched the observed experimental values [17].

This large body of work provided crucial **insights into peptide–nanoparticle binding**, as follows:

- Peptide **sequence** is important and biopanning is a powerful tool in identifying strong binders. Altering a single amino acid can lead to dramatically different binding/synthesis outcomes.
- The sequence provides a specific **structure and conformation** for the peptide, which is important in maximising interactions with surfaces.
- The flexibility/rigidity of peptide sequences is also important, as many peptides re-structure upon binding, leading to **cooperative assembly**.
- Due to the **uniqueness of peptide sequences/conformation**, they yield nanoparticles that are different from each other. This has implications for their performance (discussed below).

What we have learnt from this example, as summarised in figure 7.5, is that by learning about and utilising the nano–bio interactions, bioinspired synthesis of nanomaterials can lead to new or enhanced applications. In particular, the ability of peptides to 'recognise' different surfaces has led to the design of a range of nanocatalysts. The crucial aspect of completing this journey is gaining a molecular- or atomic-level understanding. As such, this knowledge is now being used to predictively design Pd-catalysts for a range of reactions.

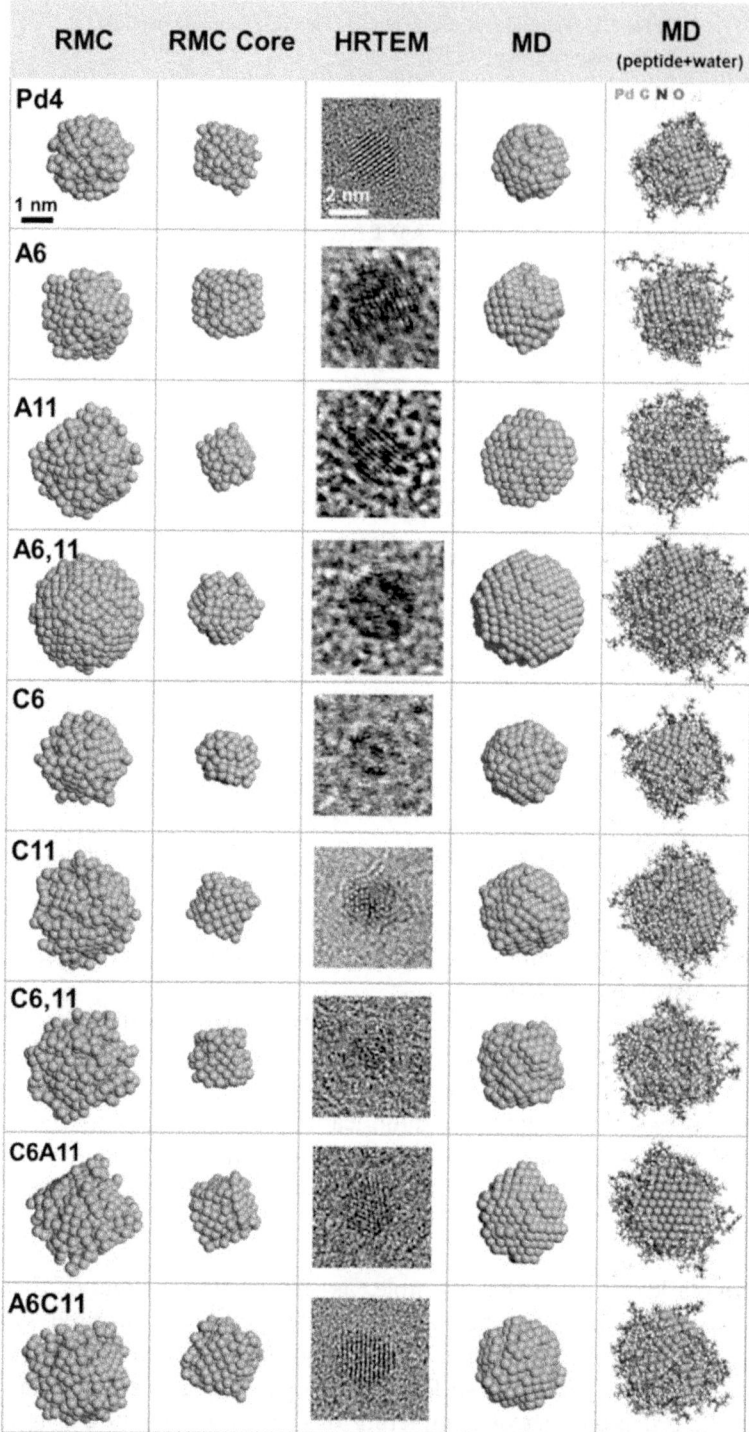

Figure 7.4. Morphologies of Pd nanoparticles synthesised using the peptides given in table 7.3. RMC = reverse Monte Carlo (for surface), RMC Core = RMC for the core of the nanoparticles, HRTEM = high resolution TEM, MD = molecular dynamics. Image taken with permission from from [17], copyright 2015 American Chemical Society.

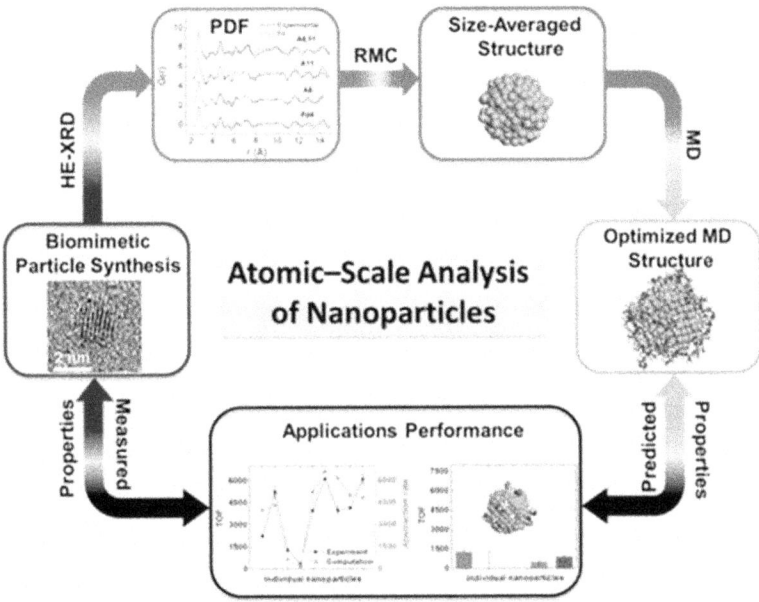

Figure 7.5. Correlation between experimentally measured reactivity of Pd nanoparticles in a Stille coupling reaction with the computed ease of surface Pd abstraction. Image taken with permission from [17], copyright 2015 American Chemical Society.

7.4 Additives as the mimics of biomineral forming biomolecules

7.4.1 The need for additives

The mechanistic understanding of the interactions between biomolecules and nanomaterials is a powerful basis for designing bioinspired syntheses of functional materials. Now the questions to consider in order to move forward are: how do we *apply* this knowledge, and what are the criteria for this *application*? The rationale behind taking this knowledge further is to enable a sustainable/green/economical route to existing materials, to enable the formation of new materials/chemistries that were not accessible previously, and/or to enable new functionalities for materials for specific applications. These needs imply that bioinspired methods provide cost-effective and eco-friendly alternatives, thereby enhancing their accessibility as well as sustainability. Further, the implication is that the methods are reproducible and scalable in order to achieve high quality products, at scale. Given these requirements, when moving from mechanistic understanding to *engineering* bioinspired products and processes, the use of specialised biomolecules (e.g. extracted proteins from organisms or peptides from biopanning) becomes problematic, for the reasons discussed below.

The extraction or genetic production of biomineralising biomolecules is not always viable. These procedures are quite complex, multistep, and require extensive purification. As such, only small quantities of biomolecules are available, typically only to those with specialised expertise, thereby **limiting access** for a wider academic and industrial community. This creates significant limitations associated with using biomolecule-based nanomaterials in sourcing sufficient materials for large-scale manufacture. Genetic engineering can provide ways for larger scale production of such molecules, however, the genetic toolkit again remains with a small group of experts. Given the *application* of the mechanistic knowledge gained requires a multidisciplinary approach, limited access to the specialised biomolecules is a key reason for the stifling of much needed further research and development. Another factor to consider is that the extracts are typically precious, delicate, and **unstable** when isolated from their native environments. This results in either the loss of their function (partly or entirely degrading over time), limiting their usage, or the need for a range of 'co-factors', thus making these biomolecules ineffective for use *in vitro*. These and other factors can result in the use of such biomolecules being **uneconomical and unsustainable** (and sometimes unsafe) for manufacturing. The added value to products or the benefits to a process can significantly diminish due to the issues discussed above, and the knowledge gained from biomineralisation is unable to meet its potential. It is therefore important to focus on mimics of biomolecules ('additives') that can provide the benefits that extracted biomolecules can, yet without the associated issues when it comes to translating the knowledge to the development of new materials, products and processes.

7.4.2 The design of additives and custom synthesis

A successful strategy involves the use of synthetic molecules (defined as '**additives**'), which can mimic the function of biomolecules, enabling custom synthesis of nanomaterials in a sustainable fashion. This strategy is detailed below in a generic sense, and is further expanded with examples of two case studies in the last two chapters of the book.

The general challenge is how to identify/select potentially active additives. We have seen from section 7.2 that the study of biomolecules and their interactions with nanomaterials has provided a range of rules and insights. These features include both the chemical and physical properties of the biomolecules and the materials. For example, the sequence of peptides as well as the chemical composition of the materials are important. In addition, the biomolecule conformation and co-operative assembly with the nanomaterials (or their precursors/intermediates) is crucial. This knowledge can be used to define the key features needed in an additive (**point 3** in figure 7.1) for green synthesis of nanomaterials. It is possible to design synthetic molecules which contain the key chemical functionality required for the synthesis and assembly of nanomaterials. One can also engineer simple (bio) molecules, e.g. self-assembling peptides and polymers, which form scaffolds for nanomaterials formation/assembly. These additives can exhibit one or more features that are associated with their biomolecule counterparts. It is important to understand how these additives interact with the entire process of nanomaterials formation (starting from precursors to the final materials), and their physical and chemical properties (e.g. self-assembly and ionisation). These bioinspired synthesis strategies (point 4 in figure 7.1) can draw from the significant progress made in the

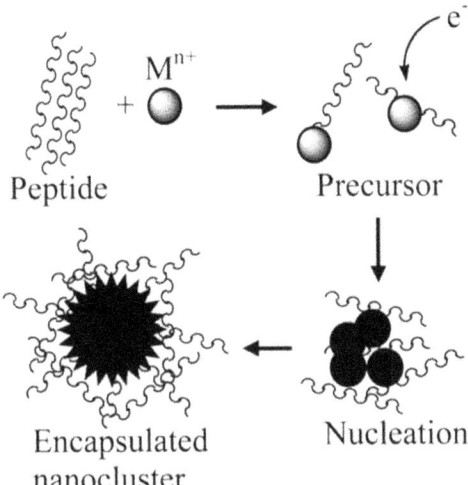

Figure 7.6. Schematic representation the roles of additives in metal nanoparticle formation. Image taken with permission from [19], copyright 2003 American Chemical Society.

synthesis of nanomaterials using bottom-up approaches, which employ sophisticated techniques such as self-assembly, nano–bio interactions, and templated synthesis [18]. These features are outlined in figure 7.6 and illustrated below with selected examples.

In section 7.2.1 and chapter 6 we have seen that some amino acids or chemical functionalities play crucial roles in biomineralisation. Examples include acidic amino acids for calcium carbonate type biominerals, and certain metal hydroxides/oxides or amines for a range of oxides and metals. One example is that of noble metal nanoparticles. A range of bacteria and fungi produce such metallic nanoparticles through the use of specialised proteins that have specific recognition against metal ions/clusters. Further, the chelation of metals with biological ligands is well known in bio-inorganic chemistry (e.g. chelation of Fe ions in heme protein). Through understanding these systems and biopanning (see section 7.2.2), it has emerged that imidazole groups of histidine and thiol groups of cysteine are key in chelating with metals. Using this knowledge, a range of synthesis strategies have been reported for metal nanoparticles (e.g. gold, silver, platinum, and copper) [19]. Using amino acids, peptides or other additives containing imidazole- or thiol-rich motifs, synthesis of these metal nanoparticles has been achieved. The roles of the additives include chelation or ion binding, followed by the formation of cluster/nuclei by lowering the surface energy, and finally controlling the growth and stabilisation of nanoparticles (see figure 7.6). In the growth stages, lattice matching and stereochemical recognition control the morphology of the final materials produced [20]. Knowing these specific (bio)chemical interactions between the additive and metals, one can develop customised bioinspired synthesis, as illustrated with examples below.

In the case of metal nanoparticle synthesis, in an effort to search for greener routes, a number of small molecules have been used, such as polyphenolic antioxidants from tea and vitamins, cellulose and winery waste [21]. For example, gold and platinum have been synthesised using vitamin B2 in water, within minutes [22]. Unlike the conventional methods described in section 4.6.5, these methods occur at room temperature in water, without the need for a separate reducing agent.

There has been extensive research reported on the design and use of a range additives in the formation of calcium-based crystals/nanoparticles [23]. These additives include biological and synthetic polymers (including copolymers), small molecules and inorganic ions. These additives are inspired by calcium carbonate biomineralisation, where proteins rich in carboxylate moieties arising from aspartic acid and glutamic acid residues have been shown to strongly bind the biomineral and regulate their formation and stability [24]. When polypeptides of these residues were investigated *in vitro*, there was clear evidence of preferential binding of these polypeptides with calcium carbonate crystals as well as their effect on crystal habits [25]. Encouraged by these results, a range of acidic additives have been investigated, such as polyacrylic acid, polystyrene sulfonate, and phytic acid (a phosphorylated sugar ring), and each have shown direct control over crystallisation. In the case of polystyrene sulfonate, for example, the polymer was found to bind free calcium ions in the solution, thus altering saturation, and hence the formation pathways. Further, the polymer was also able to preferentially bind the (011) face, causing stabilisation of this face, inhibiting its growth and leading to significant changes to the final morphology [26]. Another study using the same polymer reported that at high concentrations of polystyrene sulfonate, a less stable vaterite phase of calcium carbonate is formed instead of a more stable calcite [27].

In terms of the roles that these additives play, some areas are not well understood. For example, in calcite formation and growth, it is unclear whether the additives influence the nucleation and growth stages [23] or only the crystal growth, thus limiting control only over the morphology [28]. In addition to facilitating nucleation and growth, some additives can inhibit crystallisation via colloidal stabilisation of intermediate phases such as amorphous precursors to crystals. Indeed, in the case of calcium oxalate, the use of citrate has clearly shown that citrate chelates with the early stages of the calcium oxalate species, thus delaying or inhibiting crystallization [29]. These strategies are now being applied widely to many other crystalline nanomaterials. In chapters 8 and 9, we will look more closely at the use of designer additives in the formation of amorphous silica and crystalline magnetite.

In the above examples, the roles of additives have been limited to molecular scale interactions via charge-matching/neutralising and/or reducing the metals. Self-assembly of the additives was rarely utilised. We now focus on examples where the additive also acts as a template by assembling into supramolecular structures (see background on self-assembly in chapter 4). A classic example of this approach is the use of double stranded DNA molecules to programme the assembly of nanoparticles (figure 7.7 (top)) [30]. In this example, DNA-functionalised nanoparticles were prepared and using a complementary DNA strand(s), these nanoparticles were assembled into a desired array/aggregate. This principle has been extended to a range of biological and synthetic molecules in order to assemble nanomaterials. Another example includes the use of block co-polymers, which spontaneously

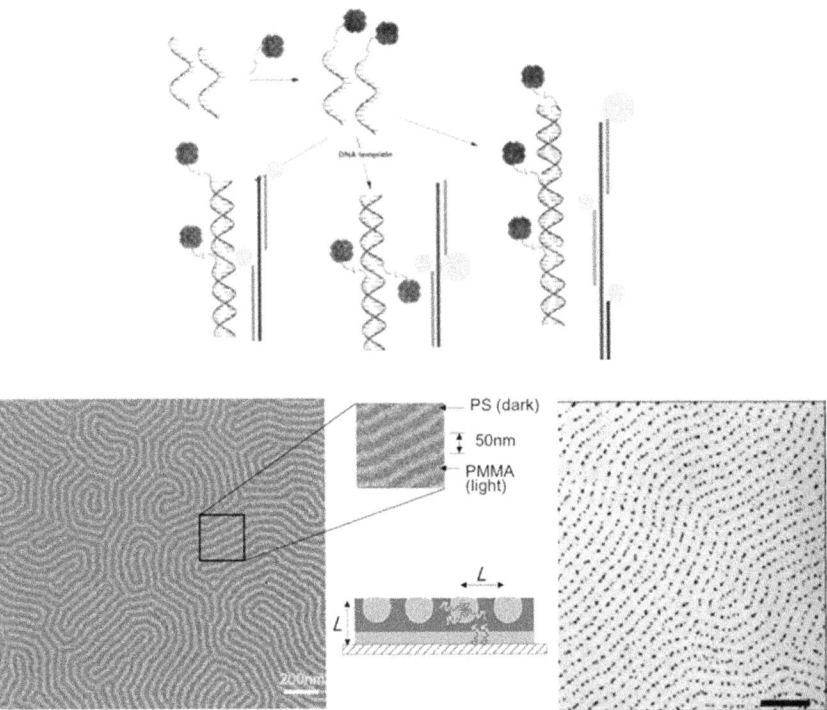

Figure 7.7. Selected examples of using chemical and physical features of (bio)molecules to assemble nanoparticles. Top: the use of DNA to programme the assembly of nanoparticles (shown as black or yellow circles). Image adapted from [30], with permission from Springer, copyright 1996. Bottom: the use of self-assembling polystyrene-b-polymethyl methacrylate (PS-b-PMMMA) block copolymers to assemble/localise nanoparticles by utilising selective chemical wetting of polystyrene block by gold nanoparticles—images show the self-assembled polymer before (left) and after (right) the addition of nanoparticles and annealing (scale bar = 200 nm). Images taken from [31], with permission from Springer, copyright 2001.

self-assemble due to localised phase separation between *immiscible* polymer blocks (figure 7.7 (bottom)) [31]. The ability of polystyrene to selectively 'wet' gold nanoparticles was exploited by using a polystyrene–poly(methyl methacrylate) co-polymer. These two blocks self-assemble, while the gold nanoparticles selectively assemble with the polystyrene block.

Another example includes the use of polyglutamic acid for the synthesis of porous alumina, where both the chemical and physical interactions are important, as depicted in figure 7.8 [32]. Under the synthetic conditions, polyglutamic acid self assembles into helices, with overall negative charge. Aluminium precursor forms cationic intermediates (hydroxide) in the aqueous synthesis system. This leads to strong ionic interactions between the additive and the inorganic species, leading to the self-assembly of aluminium hydroxide around helical polyglutamic acid. Further condensation reactions lead to the formation of alumina. A similar strategy was used in the case of $BaCO_3$ precipitation in the presence of polyelectrolytes [33]. The ability of polyelectrolytes such as poly(sodium 4-styrenesulfonate) and poly

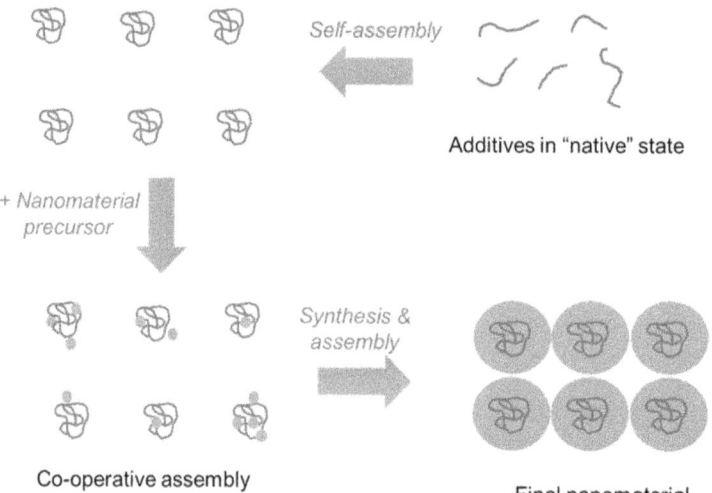

Figure 7.8. Schematic representation of how specific chemical and physical features of additives (shown in green) can be utilised in the synthesis and/or assembly of nanomaterials (shown in orange). Image adapted from [34], reproduced by permission of The Royal Society of Chemistry.

(allylamine hydrochloride) to self-assemble as well as chelate/bind mineral and/or their precursors was used to produce suprastructures of barium carbonates.

In this section, we have seen the need for more accessible additives and the benefits from using them. Using synthetic additives allows us to tailor their chemical properties to understand nanomaterials formation and design specific additives for the custom synthesis of desired nanomaterials. The knowledge of biomineralisation and the roles of organic components enable the design of such additives, which is a crucial first step towards developing sustainable bioinspired methods.

7.5 Compartmentalisation, templating and patterning

Another strategy for developing bioinspired synthesis involves the use of compartmentalisation or confinement. As seen in chapter 6, this is often the first step in biomineralisation. A compartment differentiates an area/volume to concentrate precursors, and controls the location and define the chemistry, thus controlling the material formation. Confinement or compartmentalisation offers various advantages over *homogeneous* particle formation in solutions. This strategy results in very different chemical environments compared to outside of the compartment, enabling formation and control over very specific materials. The compartment is also a template. It provides a mould for the shape and size of the final material, as well as the specific location of the material formation. Both of these factors control the material's chemical and physical features at nanoscales, while the location of the compartment can define the location of the final nanomaterial (e.g. on a surface).

The use of 'nanoreactors' (nanoscale confinement) for compartmentalisation is powerful in producing particles of precise shapes and sizes such that their aggregation can be avoided. Although synthetic nanoreactors have started to appear in the literature, most of the research to date has used biological 'compartments'. These include: within cells (e.g. magnetosomes), protein cages (ferritin), and a range of virus 'cages' [35]. Synthetic systems include self-assembled surfactant-stabilised nano-emulsions (e.g. water in oil), liposome and polymersome.

7.5.1 Confinement in a simple protein template
7.5.1.1 Ferritin
The iron storage protein ferritin has been introduced in chapter 6 (section 6.4.2.6) and is the natural place to begin when considering the biokleptic use of self-assembled biological template 'nano-reactors' for biomineralisation. This is because it is only a small re-purposing proposition. Ferritin is a simple self-assembled cage protein compartment, which already performs biomineralisation of ferrihydrite within its defined 8 nm diameter core. It is thus a simple step to use apoferritin to promote the synthesis of a range of monodisperse nanomaterials within the protein core. Ferritin is ideal for these purposes as it is inherently a biocompatible, precisely folded compartment for precision templating (of spherical particles of 6–8 nm diameter in size), with the mineralisation aided by ion channels situated at the junctions of each subunit, to enable the transport of metal ions into its core. The nanomaterials produced can be used for a wide range of applications [36]. The outer shell of ferritin can also be functionalised chemically or genetically, enabling the ferritin protein shell to act as a multivalent scaffold. For example, functionalisation of the ferritin protein can enable the particles to be soluble in organic solvents, adding more versatility to the applications for which they can be used [37, 38].

The inorganic core can be formed by directed mineral precipitation within the ferritin nanoreactor core space. The ion channels that allow iron ions to enter the core under normal conditions are generic enough to allow a full range of metal ions into the centre, from manganese to gold. The first alternative minerals to be formed within ferritin were iron sulfide and magnesium oxide in the early 90s [39], followed by magnetic nanoparticles such as magnetite [40] and CoPt [41]. Such tight control over the size and dispersity of CoPt makes very attractive particles for recording media. Indeed, the CoPt particles formed in ferritin were tested for their data recording properties [41]. Enhanced clinically applicable gadolinium (III) oxide hydroxide nanomaterials have been synthesised within ferritin [42]. Furthermore, nanotechnologically applicable cadmium sulphide quantum dots (QDs) [43], palladium [44], and gold [45] nanoparticles have also been formed within ferritin's core.

The chemistry of these systems needs to be elegantly designed. It is not simply a case of adding ferritin to a nanoparticle precipitation reaction, as this would result in a mixed population of some particle precipitating within the ferritin, but with nothing to stop particles also precipitating directly in solution. Thus, a multistep process of concentrating the ion, washing, nucleation, and then growth, can usually be adopted. For example, in the case of the Au–ferritin particles, the gold salt was included with ferritin to allow sufficient gold ions to enter the ferritin cavity (figure 7.9A). The solution was then run down a desalting column to remove the exterior gold ions, so that when the reducing agent was added only the gold ions within the ferritin could nucleate a gold nanocluster; these were then grown to fill the cavity by adding more gold salt in the presence of ascorbic acid [45]. A similar multistep process was used for the CoPt example, but in addition they also then heated the particles to anneal them as a final step to form the $L1_0$ phase of CoPt that is preferable for data storage.

7.5.1.2 Further cage protein and virus capsid templates

While ferritin is clearly applicable due to the fact it already biomineralises the nanoparticle in Nature, it is also clear that we can expand this notion to the use of a full range of self-assembled natural protein cage structures as bio–nanoreactors. MjHsp is a well-studied small and stable heat shock protein that forms a cage structure, assembled from 24 subunits with large 3 nm pores allowing reagents such as metal ions to enter the 6 nm diameter core [46]. It is stable up to 70 °C and in a pH range of 5–11, making it ideally robust for chemical synthesis conditions. MjHsp has been used to encapsulate ferrihydrite through air oxidation of Fe(II) in the presence of the protein cage, analogous to ferritin [47]. PepA is another small cage protein with a tetrahedral structure, with a 6 nm interior cavity which can be fed reagents through four 4 nm wide channels at the faces of the tetrahedron, and four 1 nm wide channels at the edges [48]. PepA has been used to mediate the growth of 1.1–2.8 nm sized CoPt magnetic nanoparticles inside the protein cavity under ambient reaction conditions, with magnetic coercivity increasing with increased particle size [48]. It is thought the positively charged metal cation precursors preferentially accumulate and nucleate in the negatively charged interior of the cage [49].

All the examples so far have had spherical cores of approximately 6 nm in size. While these are an excellent template to obtain spherical nanoparticles of this size, there is no scope for different sizes or structure with these proteins. Thus the exploitation of virus protein cages as nanoreactors for bioinspired synthesis of nanomaterials is an obvious next step. A virus is simply nucleatic acid molecules within a self-assembled protein coating/shell, which can only be replicated in a host cell using the machinery of another living organism. Once the replicated molecules in the inside are removed, the protein shells offer a family of biological structures so extensive that they offer a vast range of pre-determined compartment sizes and even a range of shapes, from very simple to ornate and complex, vastly expanding the scope for using biological cages as nanomaterial templates. Furthermore, the variety of pore sizes between the subunits offers further size-selectivity with respect to reagents entering the core and thus the nanomaterials produced.

Each virus template can be utilised in multiple ways, mineralising either within the core of the cage or even the exterior, coating the virus in a mineralised outer shell.

Internal mineralisation

Cowpea chlorotic mottle virus (CCMV) and cowpea mosaic virus (CMPV) are both icosahedral 28 nm sized viruses that infect the cowpea (black-eyed pea) plant, made up of 180 and 60 protein subunits respectively. These are both excellent examples of virus shells where the inherent properties of the structures can be utilised for

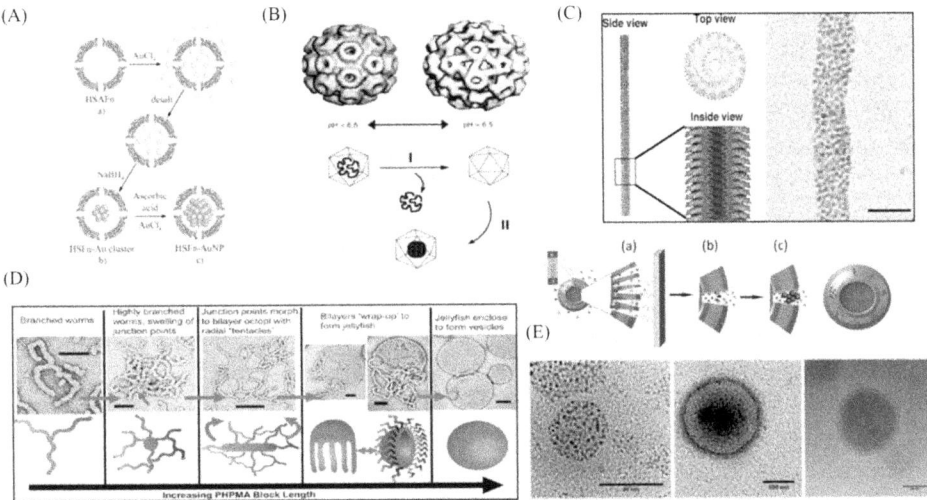

Figure 7.9. Examples of external and internal organic templates for mineralisation. (A) Use of ferritin as a nanoreactor to biomineralise precisely sized gold nanoparticles within the protein cage (reproduced with permission from [45], copyright 2010 John Wiley & Sons). (B) Example of a virus capsid that could be used as a nanoreactor for similar biomineralisation and shows the expansion of cowpea chlorotic mottle virus (CCMV) using pH conditions, demonstrating how this can aid mineralisation reactions (reproduced with permission from [52]). (C) Example of a different shaped template. Cryo-electron micrograph and reconstruction of tobacco mosaic virus (TMV) are shown with its elongated structure, and its use as template for the mineralisation of Pt nanoparticles. Scale bars 20 nm (reproduced with permission from [58]). (D) Schematic and electron microscopy interpretation of architectures achievable during polymersome synthesis demonstrating the potential for control of the formation of a polymeric nanoreactor (reprinted with permission from [71], copyright 2011 American Chemical Society). (E) Schematic and electron microscopy of the formation of magnetopolymersomes. (Ei) Schematic of how electroporation is used to open pores in the membrane to transport in soluble iron (reproduced from [78], with permission of Springer, copyright 2015). (ii) Micrographs of three examples of magnetopolymersomes: (ii) PDB–PEO magnetopolymersomes (reproduced with permission from [78], with permission of Springer, copyright 2015); (iii) modified asymmetric (carboxylate interior) PEG–PHPMA/PHMPA–PMPC magnetopolymersomes (reproduced from [79]); (iv) artificial magnetosome formed from a POPC liposome (reproduced with permission from [69]).

sophisticated bioinspired nanoparticle synthesis (figure 7.9B). CCMV has an uncharged exterior, repelling nucleation on the surface, it also has intrinsic pH-dependent pores that effectively gate the ion channels, and a positively charged inner membrane which provides an ideal environment for the nucleation and precipitation of both paratungstate and decavanadate [50, 51]. The viral capsid undergoes a reversible 10% swelling above pH 6.5. This opens up multiple pores on the virus capsid which can enable the release of the viral RNA from inside, but also allows the empty capsid to uptake precursor through these channels for nanoparticle synthesis (figure 7.9B) [50, 52]. The reversible pH switching of this process allows precise elegant control over uptake and thus precipitation of the nanomaterial through pH. CPMV has also been used as a nanoreactor to form gold [53] and silica [54], again taking advantage of their pH switchable ion transport. It should be emphasised that electrostatically induced, pH-controlled encapsulation and nucleation at a range of template sizes is something that would be extremely difficult to replicate synthetically. The tobacco mosaic virus (TMV) is a non-spherical rod-like structured virus that infects tobacco plants. It contains 2130 coat proteins that arrange into a helix around the RNA, forming a rod 300 nm long and 18 nm wide (with an internal diameter of 4 nm). The central channel of the virus has been used to template both $FePt_3$ and CoPt nanowires [55], as well as Ag nanoparticles and QDs along the internal cavity of the virus [56].

External mineralisation
Biomineralisation on the exterior of CPMV has been performed by modifying the surface to be negatively charged to enable the nucleation of cobalt and iron oxide, resulting in nanoshell materials of a very small size distribution [57]. Similarly, the exterior of the TMV has been used as a template for the synthesis of Pt nanoparticles forming nanoparticle decorated nanowires. Such structures have increased Pt surface area and the overall stability of the Pt nanostructures compared to more conventional particles, enhancing their catalytic ability [58] (figure 7.9C).

7.5.2 Confinement in modified cage protein templates

Confined bio–nanoreactors can also be combined with the use of biomolecules or additives (already discussed in sections 7.2 and 7.4). The functionalisation of the inner (and outer) walls of these nanoreactors with biomolecules derived from biomineral forming systems or their bioinspired counterparts (the additives) can be vital in controlling localisation as well as material properties. This provides the sites for heterogeneous nucleation, thereby reducing the barriers for particle formation. Further, these functional sites help control the chemical and physical properties of the final nanomaterials (e.g. chemical composition and particle shape/crystallinity). We have already considered a very generic chemical modification of the CPMV above (section 7.5.1.2), to add negative charge to the surface to enable nucleation to that surface. Here we will consider more specific modifications to direct specific mineral phase formations through nucleation with very specific biomolecules that direct this phase forming.

For example, exposed surface amino acids have been used to attach specific peptide sequences to the exterior of CPMV. The sequences (HNKHLPSTQPLA) and (CNAGDHANC) were identified to selectively bind to FePt and CoPt respectively from biopanning (section 7.2.2 and table 7.1), and it was found these sequences enhanced the formation of CoPt and FePt respectively on the surface of the virus [57, 59].

Modification of the protein cage can be directed to almost any amino acid site in the protein, so the peptide sequences can be specifically directed towards the exterior (as is the case in the examples above) or to be displayed in the protein core interior. For example, the heatshock protein MjHsp has been modified at the N-terminus (so displayed within the core) with a peptide sequence (KTHEIHSPLLHK) that shows specificity for the $L1_0$ phase of CoPt [60]. This genetically modified MjHsp produced enhanced 6.5 nm CoPt nanoparticles inside the protein cages with increased magnetic coercivity (showing some $L1_0$ character) [60].

There is also potential for engineering the nanoreactor walls such that they can enable active transport of the precursors into the reactors. Recently, this was demonstrated where the proteins responsible for active transport of silicon into diatoms were tethered on the surface of nanoreactors, which pumped in the precursor, while the 'additives' present inside the reactor facilitated the formation of silica particles [61].

7.5.3 Biomimetic compartmentalisation

Natural biomineralisation and biokleptic protein cage templates such as viral capsids are limited by what Nature provides. These bionanoreactors are also designed in the most part to operate under ambient conditions, so may not necessarily be adapted for the more robust reaction conditions required for some inorganic nanoparticle synthesis. Taking a bottom-up approach, artificial biological compartments such as artificial membrane vesicles can be produced with the synthetic requirements designed in, i.e. any new specification arising outside the biological parameters, be they more extreme in chemistry or differing in size. It cannot be underestimated how powerful it is to take components, and knowledge of both chemistry and Nature, to design and engineer in the exact requirements for nanoparticle mineralisation. This flexibility is much less achievable with biological components such as protein cages, whose evolution has been optimised by Nature for their specific natural role, and not the new role imposed upon them.

Liposomes
Liposomes are the obvious starting point, as they are the biological component most commonly used in Nature to aid biomineralisation through trafficking and concentrating precursors, as well as the nanoreactor itself for nano mineral formation (see section 6.4.2.1). Because of their intrinsic and extensive occurrence in Nature and particular relevance to medicine, liposomes have been extensively researched and characterised. There are a number of routes to form liposomes, including but not limited to: sonication, electroformation extrusion, inkjetting, and microfluidics [62]. These different preparation methods allow for multiple modes of encapsulation and in some cases encapsulation of multiple reagents. More complicated preparation methods can allow for the independent design of the inner and outer leaflet of a liposomal nanoreactor, such as the inverted emulsion method developed by Pautot *et al* [63]. Inverted emulsion allows for production of vesicles capable of withstanding two contrasting external and internal environments, widening the scope of reaction type that liposomal nanoreactors can be subjected to, particularly in cases of immiscible solutions [63]. Here we have only been considering smaller unilamellar liposomes, which are composed of just one bilayer membrane in the nanometer size range. However, it should be noted that biology utilises a full range of membrane structure, from giant unilamellar vesicles (in the micrometre range) to more complex multivesicular structures. It is thus worth considering that there is plenty of scope to

be inspired by biology to expand the sizes and shapes of templates used to more complex and elaborate membrane compartments. One such example of this is the use of a membrane tubulation (BAR) protein to create elongated membrane tubular structures from giant unilamellar vesicles. For example, giant vesicles containing QDs have been elongated with a BAR protein to make functional, biocompatible QD nanowires [64].

While we can design in more elaborate and elegant features into artificial liposomes, there is also much to be learnt from Nature about using them as a nanoreactor. A key aspect is how to transport the reagents (metal ions) across the membrane. There have been several strategies employed.

1. Encapsulate the metal ion precursors during liposome formation, and then rely on osmotic effects to increase diffusion to change the pH within the liposome to drive mineralisation. This technique has been used to form silver oxide [65] and iron oxide [66] nanoparticles within phospholipid vesicles, as well as CdS, ZnCdS and HgCdS [67] nanocrystal QDs within the core of liposomes, with a controlled one particle per liposome. However, interestingly, these particles do not fill the vesicle core. This is because the metal precursors available are finite and limited by the concentration which is encapsulated. Solid precipitated condensed mineral will always be more dense and thus smaller than the volume of the same quantity of metal ion in a solution.

2. In Nature liposomes have ion channel, specific metal uptake proteins, that provide a supply of metal ions so that the minerals can form and the crystals continue growing to fill the liposome. To mimic Nature, we could also adopt some sort of transport system for the metal ions into the liposome. This has been achieved by the incorporation of Ca^{2+} ionophore A23187 into liposomes and the loading of both Fe^{2+} and Ba^{2+} into preformed liposomes has been seen [68].

3. The ionophore is still biological. We can mimic the action of an ion channel by mechanically forming pores in the liposome membrane. This can be achieved using a process called electroporation. This is where a voltage is passed across a sample containing liposomes and this electric current induces the membrane to form temporary pores. This technique is readily used by molecular biologists as a means of inserting DNA into bacteria. Here the electroporation opens pores in the liposomes, to allow metals ions into the core. In this set up the core contains a basic (high pH) solution, so when the metal ions enter the nanoreactor liposome, they can precipitate as the oxide. This has been demonstrated for a system that mimics the magnetosome in magnetotactic bacteria, to make magnetite nanoparticles that now completely fill the liposome (figure 7.9 Eiv) [69].

Polymersomes

Liposomes are comprised of lipid amphiphilic molecules which can be quite constricting. As we are mimicking natural processes, there is nothing to stop us considering creating our vesicle membranes out of the full range of amphiphilic molecules available to us. We can thus tailor the vesicle functions even more specifically. The largest family of adaptable amphiphilic molecules are block co-polymers. Within this family there is an almost unlimited range of tunable block-lengths, functionalisation (including biocompatibility), hydrophobicity and thus architectures achievable [70, 71]. Polymersome synthesis can occur via numerous methods such as film rehydration, electroformation [72], RAFT synthesis [71, 73], ring opening polymerization [74], layer-by-layer polyelectrolyte formation utilising charge, and more complex methods such as shell cross-linked nanoparticles [75]. The diversity observed in the formation methods of polymersomes is reflected in the almost endless combination of properties that can be incorporated into a polymersome by careful selection and design of the building-block polymer materials. Just like the BAR tubulation example for liposomes, there are multiple, more complex structures that can be achieved with polymer self-assembly by controlled synthesis. For example, the structure that can be achieved with just one block co-polymer (poly (glycerol monomethacrylate)–poly(2-hydroxypropylmethacrylate) (PGMA–PHMPA)) varies from worms to nano jellyfish to vesicles, simply by increasing the length of the PHPMA block (figure 7.9D) [71]. Researchers have successfully incorporated biological components into polymer membranes [76], for example protein channels and enzymes to enable ion transport [77]. Furthermore, electroporation as a mechanical means of opening pores in liposome membranes also works on polymer membranes, which means the same technique can be used to import ions into polymersomes. This has been demonstrated using iron ions to form tiny (2–3 nm) magnetic iron oxide nanoparticles within the membrane itself. This seems to be due to the basic in the interior meeting the iron precursors at the membrane within the pore (figure 7.9 Ei) [78]. This does not mimic the magnetosomes as well as the previous lipid example of the same system (compare figure 7.9 Eii with Eiv). However, due to the tunability of the polymersomes, their inner surface can be modified to aid biomineralising within the core (figure 7.9 Eiii). This is discussed in more detail in chapter 8 (8.6.3) [79]. Polymersomes used as nanoreactors for bioinspired formation of inorganic nanomaterials is a relative new field of research, but owing to the almost unlimited range of possible polymersomes that can be designed and engineered, its potential to grow is certain.

7.5.4 Localisation and patterning on surfaces

It has already been noted that liposomes (section 6.4.2.1) are not only used as bio-nanoreaction vessel for mineralisation but are also key to trafficking and defining the site of mineralisation. To this regard, we can be inspired by nature to use biomolecules to position the site of mineralisation. So in this section the biomolecule has a dual (or even triplicate) role: It directions the formation of the inorganic material (as well as in the case of ferritin control it size) and also locates the materials on a surface in s specific pattern.

Ferritin (section 6.4.2.6 and 7.5.1.1) has been successfully immobilised on a soft lithographically patterned surface. Here micro-contact printing (see section 4.6.1) was used to pattern hydrophobic self-assembled monolayer strips alternating with positively charged 3-aminopropyltriethoxysilane (APTES) stripes on a silica surface. The negatively charged ferritin readily bound to the APTES electrostatically forming a surface of patterned ferritin [80]. In this study they heat their surfaces at 500°C in O_2 to remove the ferritin protein cage reveal just the iron-oxide particles patterned on the surface.

As we learnt in section 2.4.2.2, a key bottle neck for developing high density data storage for the next generation of HDD is finding a way to precisely pattern monodisperse hard magnetic nanoparticles on a surface. Bit patterned media requires each particle to hold a bit of information in an array. Towards this goal 7 nm CoPt nanoparticles were formed in apoferritin (as described in 7.5.1.1) and these were assembled on a silica surface [81]. The particles were not the $L1_0$ preferred crystal form for data storage so this assembly was also subjected to high-temperature (500°C–650°C) annealing to obtain the $L1_0$ phase, again removing the ferritin. In this study they also performed drag testing to see if magnetic information could be written and read on this biomediated surface.

We have already seen how ferritin can be modified with peptide earlier is section 7.5.1. Such modifications have been used to mediate the binding of ferritin to a specific surface. A specific binding sequence for titanium was inserted into ferritin. This modified ferritin cage still maintain the capability to biomineralize iron oxide, but was also capable of selectively binding to a titanium surface. Interestingly, with a higher affinity than the titanium-binding peptide in the absence of the ferritin scaffold [82].

Modifying a biomineralizing biomolecule to enable it to bind specifically to a surface has also been carried out with nanomaterial mediating biomolecules other than ferritin. There are many proteins and peptide that mediate the formation of different nanomaterials and conjugating these to surface material specific peptide to make a 'dual affinity' peptide or protein has proved to be a very successful technique. We will consider two examples, one of a protein that forms magnetite and one of a peptide that forms CoPt.

The protein Mms6 is a magnetite nucleation protein from magnetic bacteria which is described in detail in chapter 8. The unmodified protein has been attached to a patterned micro-contact printed self-assembled monolayer. Magnetite has been synthesised on this surface to make a microscale patterned array of magnetite nanoparticles [83]. The protein has also been modified with a Cys residue to bind directly to gold on a patterned surface, which again can produced a magnetite nanoparticle patterned surface when chemical synthesis is performed on the surface [84] and this has been patterned right down to the nanoscale [85].

The peptide KTHEIHSPLLHK has been used to modified ferritin to template CoPt in section 7.5.2 as it promotes the formation of L10 CoPt. This has been conjugated to a silica binding peptide to form a dual affinity peptide that will both bind to the silica surface and aid the formation of CoPt. [86] The peptide was micro-contact printed onto a silica surface and CoPt was mineralised in a pattern in a similar process to the examples above. Furthermore the KTHEIHSPLLHK has been modified with a Cys residue to bind directly to a patterned gold surface in the same way as the Cys-Mms6/magnetite example above. This has been achieved with patterns as narrow as 226 nm wide, showing a real potential for bioinspired data storage solution [87].

7.6 Scalability of bioinspired syntheses

We have learnt that biosynthesis of nanomaterials occurs across the full range of life forms, from micro-organisms to humans [88], using complex protein- and enzyme-systems. How these 'mild' processes can be mimicked using highly specific designer biomolecules is discussed in this chapter [89]. Indeed, to date, lab-scale biomolecule-mediated syntheses of ~50 inorganic nanomaterials have been reported [90], however, they are expensive and can produce only μg–mg quantities. Since such processes heavily depend on costly biomolecules [91], they are uneconomical to scale-up. Although these biological methods can be eco-friendly, they are expensive, inefficient and/or not scalable to industrial production. On the other hand, results from experiments and process calculations have suggested that the bioinspired routes, which use additives, are potentially scalable and cheaper when manufactured at large scale [91, 92]. This can also help reduce the manufacturing carbon footprint, thus providing a significant cost benefit. As such, we have seen how additives can be identified/selected to meet the desired function for developing bioinspired syntheses. However, the inability to translate these discoveries into economical and reproducible manufacturing is a critical barrier which needs to be addressed in the future [93].

Scale-up, often dubbed as *bucket chemistry*, is an essential stage in order to establish the viability for an industrial-scale production of nanomaterials using bioinspired synthesis [94]. However, bioinspired methods are currently limited to lab-scale batches [95] (there are no reports of larger-scale investigations [96]), hence their true potential has not been realised. It is important to highlight here that scale-up is not trivial, because the transport properties (e.g. mixing and heating) change non-linearly with the production scale [97, 98]. This means that although chemical rates/kinetics do not change with the production scale, the physical parameters, particularly transport properties, do [99]. As a result, the reaction pathways and resultant outcomes (e.g. the properties of nanomaterials) change with scale-up. In most cases it is unknown how this change takes place, i.e. the correlations between the production scale, the transport properties and product properties are unknown. Below we briefly list key steps that can be taken to scale-up bioinspired synthesis of nanomaterials.

An initial step involves a techno-economic assessment of the method, where it is initially assumed that the process will scale-up, and the cost of industrial manufacturing is estimated. Parameters such as the availability of resources and down-stream processing are also evaluated. Such analyses can help understand the benefits of bioinspired synthesis and identify any challenges in scale-up, e.g. cost–benefit ratio and efficiency of manufacturing. These challenges are better addressed at the discovery and design stage, due to the ease and cost-effectiveness compared to that at pilot or industrial scale.

Simultaneously, it is important to perform scalability evaluations. These include the experimental and computational approaches to develop the correlations between transport properties, the reaction timescales and the production scales. Established methods, which can be adopted by chemistry researchers, can help assess the scalability and develop pathways to scale-up [100]. Another way forward is to identify and optimise practical ways and equipment configurations [97]. Although some of these stages may seem too applied, the challenges faced during scalability analyses require further fundamental research with inputs from manufacturers. An example of this is the downstream processing (separation or purification of materials and effluent treatment), which is commonly overlooked. A simple centrifugation in the lab for the separation of materials from solvents could prove practically difficult or highly energy demanding at industrial scale. Similarly, washing performed at lab-scale can, at industrial scales, cause large quantities of waste, leading to unsustainable production. Part of the solution to these challenges could be alteration to the synthesis or even the additives used—this requires fundamental research. It is also important to be mindful of the fate of the additives used. Will they remain in the final materials? Will they be removed? If so, will they be discarded (causing pollution) or recycled? If recycled, is there an extra energy demand or cost for purification and reuse? Again, it is beneficial to consider these factors at an early (discovery) stage in order to enable swift and successful scale-up, leading to a rapid *lab-to-market* progress. Although there are not many (or any) examples of scale-up of bioinspired nanomaterials, we hope that this section has provided an awareness, a flavour of the challenges and a potential strategy to address them. With a case study of silica, chapter 9 illustrates how these principles can be applied to the scale-up of bioinspired synthesis.

7.7 Summary: key lessons about the journey towards bioinspired synthesis

One of the main purposes of developing bioinspired synthesis for nanomaterials is to create sustainable production technologies for desired products. In this chapter, we have learnt the principles of how to translate the knowledge of biomineralisation to designing biologically inspired routes. Important lessons learnt are listed below:

- A molecular level understanding of how biology produces high quality nanomaterials is crucial. One of the key controlling features of biomineralisation is the use of biomolecules. Hence understanding the roles that such biomolecules play in the entire process of biomineralisation is extremely important.
- These biomolecules have unique catalytic or binding sites that offer recognition (selectivity and high affinity), and their chemical properties (e.g. amino acid sequence in proteins or peculiar chemical structures, modifications or motifs) are important in this recognition. The structure and conformation of these biomolecules are also important because this leads to particular self-assembly (intra- and inter-molecular) and cooperative assembly with inorganic species.

- These features together enable controlled synthesis, assembly and/or functionalisation of nanomaterials.
- The direct use of biomolecules causes serious barriers to advancing green synthesis. It is therefore important to design 'additives' that can provide the benefits that extracted biomolecules can, yet without the associated issues when it comes to translating the knowledge to the development of new materials, products and processes.
- Confinement for nanomaterials synthesis, particularly when combined with the use of additives, can be powerful in controlling size, shape and localisation as well as material properties.
- It is important to be aware that scale-up is not trivial, because transport properties change non-linearly with the production scale. This means that the reaction pathways and resultant outcomes change with scale-up, and are typically unpredictable for new syntheses.

References

[1] Round F E, Crawford R M and Mann D G 1990 *The Diatoms: Biology & Morphology of the Genera* (Cambridge: Cambridge University Press)
Vrieling E G, Gieskes W W C and Beelen T P M 1999 *J. Phycol.* **35** 548

[2] Sumerel J L, Yang W J, Kisailus D, Weaver J C, Choi J H and Morse D E 2003 *Chem. Mater.* **15** 4804
Kroger N, Dickerson M B, Ahmad G, Cai Y, Haluska M S, Sandhage K H, Poulsen N and Sheppard V C 2006 *Angew. Chem. Int. Ed.* **45** 7239
Sewell S L and Wright D W 2006 *Chem. Mater.* **18** 3108

[3] Rawlings A E, Bramble J P and Staniland S S 2012 *Soft Matter* **8** 6675

[4] Kim I W, Collino S, Morse D E and Evans J S 2006 A crystal modulating protein from molluscan nacre that limits the growth of calcite in vitro *Cryst. Growth Des.* **6** 1078–82

[5] Patwardhan S V, Patwardhan G and Perry C C 2007 *J. Mater. Chem.* **17** 2875

[6] Care A, Bergquist P L and Sunna A 2015 *Trends Biotechnol.* **33** 259

[7] Klem M T, Willits D, Solis D J, Belcher A M, Young M and Douglas T 2005 *Adv. Funct. Mater.* **15** 1489

[8] Reiss B D, Mao C, Solis D J, Ryan K S, Thomson T and Belcher A M 2004 *Nano Lett.* **4** 1127

[9] Rawlings A E, Bramble J P, Tang A A S, Somner L A, Monnington A E, Cooke D J, McPherson M J, Tomlinson D C and Staniland S S 2015 *Chem. Sci.* **6** 5586

[10] Patwardhan S V, Emami F S, Berry R J, Jones S E, Naik R R, Deschaume O, Heinz H and Perry C C 2012 *J. Am. Chem. Soc.* **134** 6244

[11] Liang M-K, Deschaume O, Patwardhan S V and Perry C C 2011 *J. Mater. Chem.* **21** 80

[12] Cui Y, Kim S N, Jones S E, Wissler L L, Naik R R and McAlpine M C 2010 *Nano Lett.* **10** 4559

[13] Wang Y, Satyavolu N S R and Lu Y 2018 *Curr. Opin. Colloid Interface Sci.* **38** 158

[14] Limo M J, Sola-Rabada A, Boix E, Thota V, Westcott Z C, Puddu V and Perry C C 2018 *Chem. Rev.* **118** 11118

[15] Walsh T R and Knecht M R 2017 *Chem. Rev.* **117** 12641

[16] Pacardo D B, Sethi M, Jones S E, Naik R R and Knecht M R 2009 *ACS Nano* **3** 1288

[17] Bedford N M, Ramezani-Dakhel H, Slocik J M, Briggs B D, Ren Y, Frenkel A I, Petkov V, Heinz H, Naik R R and Knecht M R 2015 *ACS Nano* **9** 5082
[18] Weller M T, Overton T, Rourke J and Armstrong F A 2014 *Inorganic Chemistry* 6th edn (Oxford: Oxford University Press)
Ozin G A, Arsenault A C and Cademartiri L 2009 *Nanochemistry: A Chemical Approach to Nanomaterials* 2nd edn (Cambridge: Royal Society of Chemistry)
Whitesides G M and Grzybowski B 2002 *Science* **295** 5564
[19] Slocik J M and Wright D W 2003 *Biomacromolecules* **4** 1135
[20] Mann S, Archibald D D, Didymus J M, Douglas T, Heywood B R, Meldrum F C and Reeves N J 1993 *Science* **261** 1286
[21] Varma R S 2012 *Curr. Opin. Chem. Eng.* **1** 123
[22] Nadagouda M N and Varma R S 2006 *Green Chem.* **8** 516
[23] Xu A-W, Ma Y and Cölfen H 2007 *J. Mater. Chem.* **17** 415
[24] Lowenstam H A 1981 *Science* **211** 1126
Lowenstam H A and Weiner S 1989 *On Biomineralization* (New York: Oxford University Press)
[25] Gower L A and Tirrell D A 1998 *J. Cryst. Growth* **191** 153
Kato T, Suzuki T, Amamiya T, Irie T and Komiyama N 1998 *Supramol. Sci.* **5** 411
[26] Wang T, Cölfen H and Antonietti M 2005 *J. Am. Chem. Soc.* **127** 3246
[27] Jada A and Verraes A 2003 *Colloids Surf. A* **219** 7
[28] Kim Y-Y et al 2017 *Angew. Chem. Int. Ed.* **56** 11885
[29] Ruiz-Agudo E, Burgos-Cara A, Ruiz-Agudo C, Ibañez-Velasco A, Cölfen H and Rodriguez-Navarro C 2017 *Nat. Commun.* **8** 768
[30] Alivisatos A P, Johnsson K P, Peng X, Wilson T E, Loweth C J, Bruchez M P and Schultz P G 1996 *Nature* **382** 609
Loweth C J, Brett Caldwell W, Peng X, Paul Alivisatos A and Schultz P G 1999 *Angew. Chem. Int. Ed.* **38** 1808
[31] Lopes W A and Jaeger H M 2001 *Nature* **414** 735
[32] Jan J S and Shantz D F 2005 *Chem. Commun.* **16** 2137
[33] Yu S-H, Cölfen H, Xu A-W and Dong W 2004 *Cryst. Growth Des.* **4** 33
[34] Patwardhan S V, Mukherjee N, Steinitz-Kannan M and Clarson S J 2003 *Chem. Commun.* **10** 1122
[35] Rong J, Niu Z, Lee L A and Wang Q 2011 *Curr. Opin. Colloid Interface Sci.* **16** 441
[36] Jutz G, van Rijn P, Miranda B S and Boker A 2015 *Chem. Rev.* **115** 1653
[37] Wong K K W, Colfen H, Whilton N T, Douglas T and Mann S 1999 *J. Inorg. Biochem.* **76** 187
[38] Sengonul M, Ruzicka J, Attygalle A B and Libera M 2007 *Polymer* **48** 3632
[39] Meldrum F C, Wade V J, Nimmo D L, Heywood B R and Mann S 1991 *Nature* **349** 684
[40] Meldrum F C, Heywood B R and Mann S 1992 *Science* **257** 522
[41] Mayes E, Bewick A, Gleeson D, Hoinville J, Jones R, Kasyutich O, Nartowski A, Warne B, Wiggins J and Wong K K W 2003 *IEEE Trans. Magn.* **39** 624
[42] Sanchez P, Valero E, Galvez N, Dominguez-Vera J M, Marinone M, Poletti G, Corti M and Lascialfari A 2009 *Dalton Trans.* **5** 800
[43] Wong K K W and Mann S 1996 *Adv. Mater.* **8** 928
[44] Ueno T, Suzuki M, Goto T, Matsumoto T, Nagayama K and Watanabe Y 2004 *Angew. Chem.* **116** 2581

[45] Fan R, Chew S W, Cheong V V and Orner B P 2010 *Small* **6** 1483
[46] Kim K K, Kim R and Kim S-H 1998 *Nature* **394** 595
[47] Flenniken M L, Willits D A, Brumfield S, Young M J and Douglas T 2003 *ACS Nano Lett.* **3** 1573
[48] San B H, Lee S, Moh S H, Park J-G, Lee J H, Hwang H-Y and Kim K K 2013 *J. Mater. Chem. B* **1** 1453
[49] Douglas T, Strable E, Willits D, Aitouchen A, Libera M and Young M 2002 *Adv. Mater.* **14** 415
[50] Douglas T and Young M 1998 *Nature* **393** 152
[51] Johnson J E and Speir J A 1997 *J. Mol. Biol.* **269** 665
[52] Speir J A, Munshi S, Wang G, Baker T S and Johnson J E 1995 *Structure* **3** 63
[53] Aljabali A A A, Lomonossoff G P and Evans D J 2011 *Biomacromolecules* **12** 2723
[54] Steinmetz N F, Shah S N, Barclay J E, Rallapalli G, Lomonossoff G P and Evans D 2009 *Small* **5** 813
[55] Tsukamoto R, Muraoka M, Seki M, Tabata H and Yamashita I 2007 *Chem. Mater.* **19** 2389
[56] Dujardin E, Peet C, Stubbs G, Culver J N and Mann S 2003 *Nano Lett.* **3** 413
[57] Aljabali A A A, Barclay J E, Cespedes O, Rashid A, Staniland S S, Lomonossoff G P and Evans D J 2011 *Adv. Funct. Mater.* **21** 4137
[58] Górzny M Ł, Walton A S, Wnęk M, Stockley P G and Evans S D 2008 *Nanotechnology* **19** 165704
Górzny M Ł, Walton A S and Evans S D 2010 *Adv. Funct. Mater.* **20** 1295
[59] Aljabali A A A, Shah S N, Evans-Gowing R, Lomonossoff G P and Evans D J 2011 *Integr. Biol.* **3** 119
[60] Klem M T, Willits D, Solis D J, Belcher A M, Young M and Douglas T 2005 *Adv. Funct. Mater.* **15** 1489
[61] Knight M J, Senior L, Nancolas B, Ratcliffe S and Curnow P 2016 *Nat. Commun.* **7** 11926
[62] Immordino M L, Dosio F and Cattel L 2006 *Int. J. Nanomed.* **1** 297
Maherani B, Arab-Tehrany E, Mozafari M R, Gaiani C and Linder M 2011 *Curr. Nanosci.* **7** 436
Walde P, Cosentino K, Engel H and Stano P 2010 *Chembiochem.* **11** 848
[63] Pautot S, Frisken B J and Weitz D A 2003 *PNAS* **100** 10718
Pautot S, Frisken B J and Weitz D A 2003 *Langmuir* **19** 2870
[64] Tanaka M, Critchley K, Matsunaga T, Evans S D and Staniland S S 2012 *Small* **8** 1590
[65] Mann S and Williams R J P 1983 *J. Chem. Soc., Dalton Trans.* **2** 311
[66] Mann S, Hannington J P and Williams R J P 1986 *Nature* **324** 565
[67] Kennedy M T, Korgel B A, Monbouquette H G and Zasadzinski J A 1998 *Chem. Mater.* **10** 2116
Korgel B A and Monbouquette H G 2000 *Langmuir* **16** 3588
[68] Chakrabarti A C, Veiro J A, Wong N S, Wheeler J J and Cullis P R 1992 *Biochim. Biophys. Acta Biomembr.* **1108** 233
[69] Bakhshi P K, Bain J, Gul M O, Stride E, Edirisinghe M and Staniland S S 2016 *Macromol. Biosci.* **16** 1555–61
[70] Mai Y and Eisenberg A 2012 *Acc. Chem. Res.* **45** 1657–66
[71] Blanazs A, Madsen J, Battaglia G, Ryan A J and Armes S P 2011 *J. Am. Chem. Soc.* **133** 16581–7

[72] Discher B M, Won Y-Y, Ege D S, Lee J C-M, Bates F S, Discher D E and Hammer D A 1999 *Science* **284** 1143
Discher D E and Eisenberg A 2002 *Science* **297** 967
[73] Blanazs A, Armes S P and Ryan A J 2009 *Macromol. Rapid Commun.* **30** 267
[74] Zhang F, Smolen J A, Zhang S, Li R, Shah P N, Cho S, Wang H, Raymond J E, Cannon C L and Wooley K L 2015 *Nanoscale* **7** 2265
Lutz J-F 2010 *Nat. Chem.* **2** 84
[75] Zhang Q, Remsen E E and Wooley K L 2000 *J. Am. Chem. Soc.* **122** 3642
[76] Meier W, Nardin C and Winterhalter M 2000 *Angew. Chem. Int. Ed.* **39** 4599
[77] Onaca O, Nallani M, Ihle S, Schenk A and Schwaneberg U 2006 *Biotechnol. J.* **1** 795
[78] Bain J, Ruiz-Perez L, Kennerley A J, Muench S P, Thompson R, Battaglia G and Staniland S S 2015 *Sci. Rep.* **5** 14311
[79] Bain J, Legge C J, Beattie D L, Sahota A, Dirks C, Lovett J R and Staniland S S 2019 *Nanoscale* **11** 11617
[80] Martinez R V, Chiesa M and Garcia R 2011 Nanopatterning of ferritin molecules and the controlled size reduction of their magnetic cores *Small* **7** 2914–20
[81] Mayes E, Bewick A, Gleeson D, Hoinville J, Jones R, Kasyutich O, Nartowski A, Warne B, Wiggins J and Wong K K W 2003 Biologically derived nanomagnets in self-organized patterned media *IEEE Transactions on Magnetics* **39** 624–7
[82] Sano K-I, Ajima K, Iwahori K, Yudasaka M, Iijima S, Yamashita I and Shiba K 2005 Endowing a ferritin-like cage protein with high affinity and selectivity for certain inorganic materials *Small* **1** 826–32
[83] Galloway J M, Bramble J P, Rawlings A E, Burnell G, Evans S D and Staniland S S 2012 Biotemplated magnetic nanoparticle arrays *Small* **8** 204–8
[84] Bird S M, Galloway J M, Rawlings A E, Bramble J P and Staniland S S 2015 Taking a hard line with biotemplating: cobalt-doped magnetite magnetic nanoparticle arrays *Nanoscale* **7** 7340–51
[85] Bird S M, El-Zubir O, Rawlings A E, Leggett G J and Staniland S S 2016 A novel design strategy for nanoparticles on nanopatterns: interferometric lithographic patterning of Mms6 biotemplated magnetic nanoparticles *J. Mater. Chem. C*
[86] Galloway J M, Talbot J E, Critchley K, Miles J J and Bramble J P 2015 Developing biotemplated data storage: room temperature biomineralization of L10 CoPt magnetic nanoparticles *Adv Funct Mater* **25** 4590–600
[87] Galloway J M, Bird S M, Talbot J E, Shepley P M, Bradley R C, El-Zubir O, Allwood D A, Leggett G J, Miles J J and Staniland S S *et al* 2016 Nano- and micro-patterning biotemplated magnetic CoPt arrays. *Nanoscale* **8** 11738–47
[88] Staniland S, Williams W, Telling N, Van der Laan G, Harrison A and Ward B 2008 *Nat. Nanotechnol.* **3** 158
[89] Galloway J M, Bramble J P and Staniland S S 2013 *Chem. Eur. J.* **19** 8710
Patwardhan S V, Emami F S, Berry R J, Jones S E, Naik R R, Deschaume O, Heinz H and Perry C C 2012 *J. Am. Chem. Soc.* **134** 6244
Rawlings A E, Bramble J P, Tang A A S, Somner L A, Monnington A E, Cooke D J, McPherson M J, Tomlinson D C and Staniland S S 2015 *Chem. Sci.* **6** 5586
[90] Patwardhan S V, Patwardhan G and Perry C C 2007 *J. Mater. Chem.* **17** 2875
Dickerson M B, Sandhage K H and Naik R R 2008 *Chem. Rev.* **108** 4935
[91] Drummond C, McCann R and Patwardhan S V 2014 *Chem. Eng. J.* **244** 483

[92] Patwardhan S V and Perry C C 2010 *Silicon* **2** 33
[93] Roco M C, Mirkin C A and Hersam M C 2010 Nanotechnology research directions for societal needs in 2020: retrospective and outlook *Panel Report* (Lancaster, PA: WTEC) https://www.nano.gov/sites/default/files/pub_resource/wtec_nano2_report.pdf
Patwardhan S V, Manning J R H and Chiacchia M 2018 *Curr. Opin. Green Sustain. Chem.* **12** 110–6
[94] Van Gerven T and Stankiewicz A 2009 *Ind. Eng. Chem. Res.* **48** 2465
[95] Dahl J A, Maddux B L S and Hutchison J E 2007 *Chem. Rev.* **107** 2228
[96] Kharissova O V, Dias H V R, Kharisov B I, Perez B O and Perez V M J 2013 *Trends Biotechnol.* **31** 240
[97] Yates J A 1997 *Pilot Plants and Scale-up of Chemical Processes* vol 1 ed W Hoyle (Cambridge: The Royal Society of Chemistry), p 3
[98] Sharratt P N 1997 *Pilot Plants and Scale-up of Chemical Processes* vol 1 ed W Hoyle (Cambridge: The Royal Society of Chemistry), p 13
[99] Atherton J H 1999 *Pilot Plants and Scale-up of Chemical Processes* vol 2 ed W Hoyle (Cambridge: The Royal Society of Chemistry), p 19
[100] Bourne J R and Bałdyga J 1999 *Turbulent Mixing & Chemical Reactions* (New York: Wiley)
Marchisio D L, Omegna F, Barresi A A and Bowen P 2008 *Ind. Eng. Chem. Res.* **47** 7202
Bourne J R 2003 *Org. Process Res. Dev.* **7** 471

Section IV

Case studies

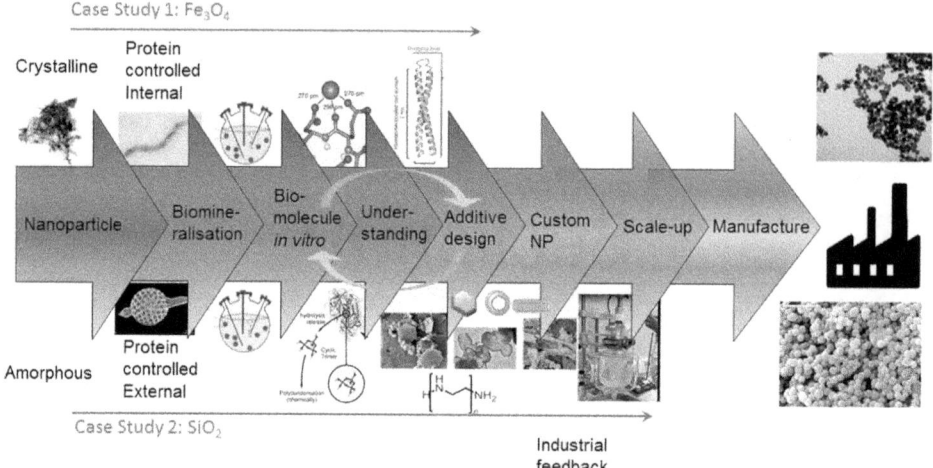

Through the course of this book we have learnt what nanomaterials are, and the real need for them in many varied applications in modern and future contexts. We have explored traditional methods of making nanomaterials, and then compared these to those in Nature. This book has then asked you, the reader, to seek inspiration from Nature, to use its tools and ingenious methods to develop green bioinspired methods of producing nanomaterials. This is beneficial with respect to reduced waste and toxicity of waste, but also goes beyond environmental advantages, potentially providing cheaper methods to create more precise and tailored nanomaterials. It is a learning pipeline, as depicted above. Exploring and learning the fundamentals and concepts behind how Nature makes biominerals in biomineralisation, and how we can understand this process and use them in bioinspired designed synthesis, are covered in section 2, which builds up in steps along this pipeline. However, clarity is often gained, and learning cemented, with an example or two. These examples will run through the processes presented, which are more generally described in chapters 5–7.

The final two chapters plot the journey of a specific material (one for each chapter) through the principles identified throughout this book, from biomineralisation to bioinspired green synthesised nanomaterial manufacturing, to offer this clarity. The two examples chosen are magnetite nanoparticles and nanoporous silica. They are selected to offer contrasting examples, as they are quite different in several ways. Magnetite is a nano-particular crystalline, well-ordered material, whereas silica is nano-porous and amorphous, lacking any order at the atomic level. Additionally, magnetite is biomineralised internally within magnetic bacteria in nature, whereas silica is biomineralised externally on the walls of diatoms. While both are a synthesised chemical through precipitation techniques, the chemistry is also different. Furthermore, industrially, silica is one of the most widely used materials in applications, with multimillions of tonnes manufactured commercially, whereas a market for precise nanoparticulate magnetite is just emerging with the advent of nanomedicine. As such, both are at different stages of development on the manufacturing journey. In case study 1, magnetite is only just beginning its journey down this bioinspired synthesis pipeline, and research is expected to increase in intensity in the later steps in the future. In contrast, bioinspired silica is further along the process, with much significant understanding already in place to begin exploring scale and manufacture sooner. We will start with magnetite, where the emphasis is on bioinspired production at the lab scale, not having progressed as far down the pipeline as silica. This case study will cover the scientific understanding of the biomineralisation process in depth, and how a bioinspired additive can be developed from this understanding, with a future perspective of where this might lead to with respect to manufacturing. The second case study on silica demonstrates how much further along the journey it is. Research on biosilica, bioinspired silica and synthetic silica has seen tremendous advances recently, and hence this case study extends further into the realms of scale-up and provides a strong guide to developing green manufacturing routes for nanomaterials in the future.

What both examples have in common can be clearly seen from these case studies: we can learn from the natural process, understand the natural process, develop/design and understand the natural process mixed with a synthetic process in the form of bioinspired green nanomaterial synthesis, and develop these additive and methods for scale-up and manufacture.

IOP Publishing

Green Nanomaterials
From bioinspired synthesis to sustainable manufacturing of inorganic nanomaterials
Siddharth V Patwardhan and Sarah S Staniland

Chapter 8

Case study 1: magnetite nanoparticles

8.1 Magnetite biomineralisation in magnetotactic bacteria

Magnetotactic bacteria (figure 8.1(I)) were first identified as early as 1963 by an Italian student who reported their strange magnetotactic behaviour in his thesis [1]. They were then rediscovered and more widely reported by Blakemore in 1975 [2]. While there have been reports of a range of magnetic bacteria, archaea, and multicellular microbes [3], the majority of them fall into the category of a gram-negative aquatic prokaryotic microbe that uses flagellae to propel themselves around the waterways they inhabit. They tend to be microaerobic and thus inhabit the very specific oxic–anoxic transition zone (where oxygen in the water depletes as the depth increases) in their aquatic environments, where an oxygen concentration of about 1% is optimum for them to survive [3]. This does unfortunately make them very difficult for researchers to cultivate and manipulate in the lab. Magnetotactic bacteria biomineralise magnetite [4], although some more exotic strains do also biomineralise greigite [5, 6]. As the ability to biomineralise nanomagnets has now been identified across a vast range of micro-organisms, it has been proposed that the molecular biological machinery required to biomineralise nanomagnets is transferred through horizontal gene transfer. However, there is also another hypothesis that it is a very ancient trait, with all strains originating from a common ancestor from when the Earth's atmosphere was more iron-rich, and had lower oxygen concentrations. In this theory, the scattered microbes that biomineralise nano-magnets are the only ones remaining with this ancient ability, while the rest of the microbes have lost the genes required for biomineralisation over millenia [7].

Magnetotactic bacteria possess liposomes within their cells called magnetosomes, in which the magnetite is biomineralised. Each magnetosome consists of a lipid membrane containing biomineralisation proteins, surrounding a magnetite nanoparticle of precise morphology dictated by the bacterial strain (figure 8.1(II)) [8–10]. The nanoparticles are in a size range of 35–120 nm in diameter, which coincides with the range for single domain magnetism for magnetite, giving them the maximum

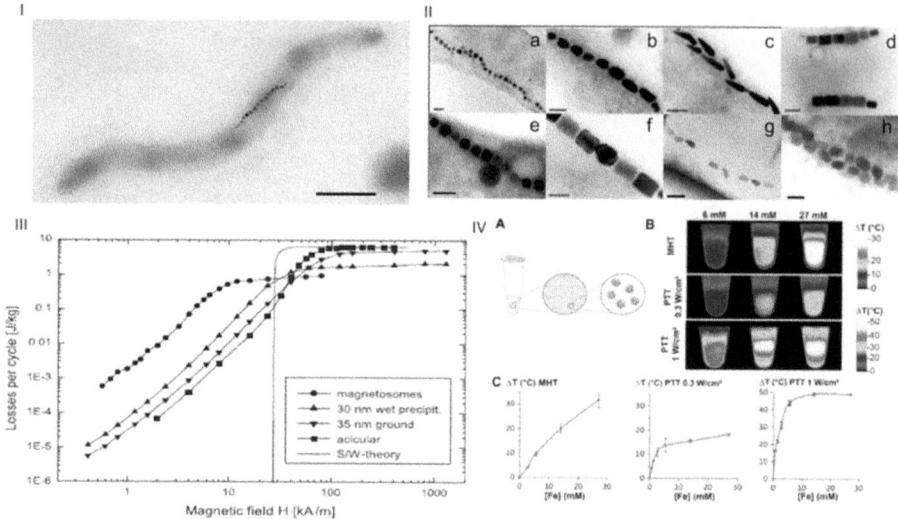

Figure 8.1. (I) TEM micrograph of a magnetotactic bacterium *Magnetospirillum magnetotacticum* MS-1 (scale bar 1 μm), reproduced from [24], copyright 2009 John Wiley & Sons. (II) TEM images of magnetosomes from different strains of magnetotactic bacteria, showing various morphologies: (a)–(g) magnetite particles and (h) greigite particles. (a) and (e) Show cubo-octahedral magnetosomes, (a) from *M. gryphiswaldense* MSR 1, (e) from *M. magnetotacticum* MS-1. (b) Bullet-shaped magnetosomes and (c) tooth-shaped magnetosomes. (d) and (f) Are elongated magnetosomes from (d) *M. coccus* MC 1 and (f) *M. vibros* MV-1. (g) and (h) Irregular bullet-shaped magnetosomes from (g) *D. magneticus* RS 1 and (h) uncharacterised greigite-producing bacterium. Scale bars are 100 nm. Reproduced from [24], with kind permission from D Schüler and T Matsunaga for some of the images. (III) Graph showing dependence of magnetic hysteresis heat loss on the amplitude of the magnetic field for a range of different single-domain magnetite nanoparticles and the theoretical value. Reprinted from [22], copyright IOP Publishing. Reproduced with permission. All rights reserved. (IV) Heating efficiency of magnetosome in water. (A) Schematic of the particle suspension. (B) IR images of magnetosome suspensions at three different concentrations generating heat through magnetic hyperthermia (MHT) and two different powers of photothermia (PTT). (C) Graphical representation of (B). Reproduced from [23], copyright 2018 with permission from Elsevier.

magnetic moment possible [11]. There are typically between 20 and 40 magnetosomes per cell, and they are arranged in long chains, which behave like a compass needle inside the bacteria, allowing them to respond to a magnetic field. In Nature this means that they align to the Earth's magnetic field [12]. A response to a magnetic field is called magnetotaxis [13, 14]. The reason why magnetotactic bacteria display magnetotaxis is not understood. It is widely believed that they use it for orientation to navigate (possibly using a mixture of aerotaxis and magnetotaxis), as they live most optimally in a very narrow chemically defined region in the aquatic environment. However, other functions have been suggested, such as detoxification, catalysis and metabolism. It is not straightforward, and no one hypothesis has yet emerged as the most likely.

8.2 Magnetosome use in applications: advantages and drawbacks

Advantages

Magnetosomes offer an excellent nanomaterial for multiple applications. They are high crystalline, are extremely homogeneous, and are coated in an intrinsic biological lipid membrane, which can be readily functioned and decorated by attaching drugs or biomarkers, etc, making them exceptionally attractive for a range of therapeutic and diagnostic biomedical applications [15].

The lipid coating also means that magnetosomes disperse exceptionally well in aqueous solution, meaning they do not clump or aggregate during biomedical use. Similar quality synthetically produced particles need to be coated post-production in a lipid to bring about similar properties.

Iron oxide magnetic nanoparticles can be used as a carriers for drugs, DNA, proteins or other relevant molecules to facilitate magnetically targeted treatments. For magnetosomes this is fairly easy, as the majority of the lipids in the membrane are amine terminated, allowing easy attachment of biotin, which can then be used to couple to streptavidin, which can in turn be linked to the functional molecule [16]. Another approach to functionalisation is to make genetically engineered magnetotactic bacteria to produce biotinylated proteins anchored to the magnetite nanoparticles, displayed on the surface of the magnetosome membrane, which can then be tagged with streptavidin linked molecules, as before [17]. These methods are in addition to direct covalent attachment to the lipid membrane. The magnetite nanoparticles can then be magnetically guided to a particular site for localized drug treatment or gene therapy, as outlined in chapter 2.

In one study, doxorubicin (DOX), a well-known anti-cancer agent, was attached to magnetosomes and applied to mice bearing a tumour [18]. The attached magnetosome allows the drug to be guided to and accumulated at the site of the tumour, giving a high tumour kill rate with reduced side effects. Such targeted treatments allow gene or drug delivery to be carefully directed to the best location for a successful outcome, and often means lower doses can be used.

Magnetosomes also display superior properties when utilised as contrast agents in magnetic resonance imaging (MRI). The high magnetization of magnetosomes over synthetic nanoparticles gives improved contrast, allowing a more detailed picture of disease states to be observed [19]. Similarly, magnetosomes show exceptional properties with respect to hyperthermic therapies (where an alternating magnetic field is used to heat the magnetic nanoparticles). Magnetosomes have the largest specific heating power compared to comparable synthetic magnetite nanoparticles (up to 5× greater!), making them well suited to this promising treatment area (figure 8.1(III)) [20–22]. Although it is not clear why this is the case, it has been proposed that it is due to the high degree of crystallinity within magnetosomes, or conversely to the subtly different, slightly reduced form of magnetite the magnetosomes produce, which cannot be emulated synthetically. Furthermore, magnetosomes have recently shown exceptional promise as nanomaterials for photothermic therapies, showing even better heating than magnetic hyperthermia (figure 8.1(IV)) [23].

Disadvantages

While magnetosomes offer highly desirable qualities for biomedical applications, they are not currently used commercially, as there are two main drawbacks of the biosynthetic method. The first is that the slow growth rate and low growth density of magnetic bacterial cells give low yields of magnetosomes. Several research groups have developed improved fermentation methods with improved yields (0.32 g l^{-1} day^{-1}, giving a magnetosome yield of 39 mg g^{-1} dry cells), by optimizing media nutrient and oxygen concentration [25]. This has been improved further simply by increasing the air aeration/stirring rate, resulting in 2.17 g l^{-1} at a growth rate of 0.868 g l^{-1} day^{-1}, giving a magnetosome yield 2.5 times higher than the previous result [26]. While these are clear improvements, biomineralisation in magnetotactic bacteria, even in an improved fermentation system, is still an inefficient method of producing magnetic nanoparticles compared to other commercial methods, presenting a real barrier to its successfully commercialisation.

The second drawback of the biosynthesis of magnetosomes within cells is the lack of flexibility this methodology offers. Magnetosomes are almost exclusively pure magnetite of highly specific sizes and shapes, due to the high degree of control exerted by the cell. This is in many ways an asset, if this is exactly what is required. However, if you do not require this exact specification of nanoparticle, the natural system offers little. Thus the applicability of natural unaltered magnetosomes is limited to a small range of magnetite materials of very specific shapes and sizes. This is beginning to be addressed: doping the magnetite magnetosomes with other transition metals such as cobalt is possible *in vivo*, simply by adding these metal ions into the growth media [27]. However, only small quantities can be added to the nanocrystal before the non-ferrous metals have a toxic effect. This also means that growth is retarded and thus yield is reduced. The maximum doping of cobalt into magnetosome has been approximately 3% doping (metal content) [28]. Both manganese and copper have also been doped into magnetosomes, with the former showing less uptake, and the latter having increased uptake but also increased toxicity [28]. The cobalt-doped magnetosomes have increased coercivity, which in turn increased hyperthermic heating in mice [29]. However, the route to flexibility is limited by the metals Nature will accommodate (so far only three transition metal dopants achieved) and concentrations are restricted by the organism. Furthermore, these doped particles are obtained in even lower yields due to the poisoning effect of the dopant, reducing the commercial viability further. Thus, it is prudent to also consider a bioinspired approach to making magnetosomes and magnetosome-like magnetite nanoparticles.

8.3 Biomolecules and components controlling magnetosome formation

8.3.1 Magnetosome biomineralisation protein discovery

The previous section highlights many of the advantages of biosynthesis, but also the disadvantages relating to production and flexibly around adapting the material, meaning the route is not currently accessible for scale-up and manufacture. Thus, we seek inspiration from Nature to aim to produce nanoparticles with the best of both worlds: biosynthetic precision, under green ambient formation conditions, but at high yields, with the ability to design and tailor the particles. In order to do this, we must first understand the biomineralisation process and thus discover which biological components are controlling it.

The names of the magnetosome biomineralisation proteins have mainly been annotated as Mam (magnetosome associated membrane) protein (followed by a letter), led by the Schuler research group working on the MSR-1 strain, whereas the Matsunaga group used the nomenclature of Mms (magnetosome membrane specific) proteins (followed by a number denoting the protein size in K Daltons, or a letter) for the AMB-1 strain. Due to the different groups working on similar proteins concurrently (their similarity unknown at the time), several homologous proteins have the same name. E.g., Mms13 in AMB-1 is homologous to MamC in MSR-1. There are many Mam/Mms proteins, where homologous proteins are found across magnetosome producing bacteria and are unique to only these bacteria. These are clearly good magnetosomes biomineralisation candidates to investigate.

8.3.1.1 Magnetosome genes: the magnetosome island
Interestingly, the genes that encode the Mms and Mam proteins unique to magnetic bacteria are found in one region of the magnetotactic bacterial genome (approximately 100 kilobase). This has been called the magnetosome island (MAI), and encodes around 100 proteins. Genes which encode proteins with similar functions, involved in similar processes, or co-dependent on one another, are often found together in the genome in regions called clusters. Major clusters in the MAI are the mamAB cluster, which encodes 17 genes for the production of proteins for iron transport, magnetite formation, and protein sorting, and which on its own is capable of producing a minimal magnetosome [30]. However, this is not equivalent to native magnetosomes. There has been a detailed knock-out mutagenesis (Δ) study on this gene cluster, where a gene of choice is disrupted so the magnetic bacteria no longer produce the protein that gene encodes. If the cells now fail to make magnetosomes, or the quality of the magnetosomes is very poor, it shows the protein has a role in magnetosome formation. A systematic gene knock-out survey has been performed where each gene was deleted in turn to see which were the most important genes for magnetite production (figure 8.2), identifying critical genes and thus proteins [31]. These will be discussed in more detail in later sections [31]. Later research found that four gene clusters from the MAI (along with a couple of gene clusters from elsewhere in the genome) are required to form magnetosomes within different non-magnetic bacterial host cells. These four clusters are mamAB (already mentioned), as well as the mamGDFC cluster (which encodes four genes), the mamXYZ cluster (of three genes), and the mms6 gene cluster (containing five genes). The other two gene clusters are FeoAB1, which contains genes that encode for iron transporter proteins, and an FtsZm cell division gene cluster (figure 8.2) [32]. This research, for the first time produced magnetosomes in a non-magnetic bacterial strain, by inserting these genes into *R. rubrum* bacteria, giving under 35 genes that are necessary/critical for

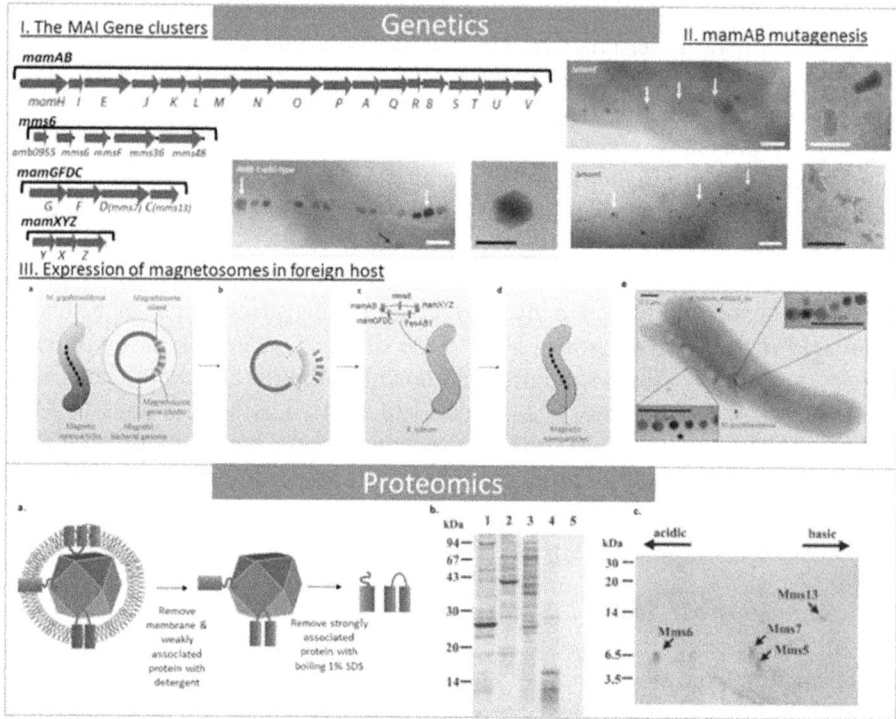

Figure 8.2. Demonstrating genetic (top) and proteomic (bottom) methods of biomineralisation protein discovery for magnetic bacteria. Genetic: (I) shows the four essential magnetosome island gene clusters; (II) shows images from a knock-out mutagenesis study of the individual genes from the mamAB cluster (top ΔmamT; bottom ΔmamS). The central image is the control wild type *Magnetospirillum magneticum* AMB-1. Scale bars are 100 nm for full chain images and 50 nm for close up of the magnetosomes. Image reproduced from [31]. (III) (a)–(d) Schematic showing how the four essential gene clusters (shown in (I)) along with an iron transport cluster (FeoAB1), and FtsZm gene, can be expressed in a non-magnetic bacterium (*R. rubrum*) to enable the production of magnetosomes in a foreign host. Reproduced from [33] by permission from Macmillan Publishers Ltd. (e) TEM image of magnetosome bearing *R. rubrum*, reproduced from [32] by permission from Macmillan Publishers Ltd. Proteomics: (a) schematic showing how the magnetosome membrane containing many proteins can be removed with detergent, leaving only proteins that are strongly bound to the magnetite crystal, which can then be removed for analysis with 1% boiling sodium dodecyl sulfate (SDS). (B) A gel showing the protein masses of *M. magneticum* AMB-1 protein extracted from lane 1. The bacterial magnetosomes: lane 2. The cell membrane: lane 3. The cell cytoplasm: lane 4. The proteins tightly bound to the magnetite crystal: lane 5, show there is no protein inside the magnetite crystal itself. (C) 2D electrophoresis gel of the fraction shown in lane 4 (tightly associated magnetosome proteins). Both (b) and (c) reproduced with permission from [34].

magnetosome production in an alternative host organism [32]. Genetic analysis of this kind allows researchers to recognise the critical genes and thus the proteins they encode for further study and understanding.

8.3.1.2 Magnetosome proteomics
Initially, a proteomics approach was used to identify the proteins specific to the magnetosome membrane, and as such both genetic and proteomic research was

performed in parallel. In this process, the bacteria were lysed and the magnetosomes magnetically separated from the rest of the cell. The membrane was then removed with detergents, and the resulting protein mixture then separated and analysed. Some proteins were only loosely associated with the membrane, while others at the other extreme were tightly bound to the magnetite crystal (figure 8.2). Based on sequence homology, many of the proteins belong to well-characterised families, which give a good indication of the likely role they play in the bacteria. However, several of the proteins are unique to magnetotactic bacteria, with no known similar proteins to give indications of function. It is thus likely that these proteins interact to biomineralise the nanoparticle, performing functions such as particle nucleation, maturation, and directing the particles' crystal growth. The function of Mms/Mam biomineralisation proteins found through proteomic routes can then be probed through knock-out mutagenesis, and a detailed analysis of the protein sequence can be performed; comparison to known proteins with known functions can help to understand the function of the new protein.

8.3.2 Bio-components for each step of biomineralisation

Within the magnetosome, proteins control all aspects of magnetite formation. The biomineralisation of magnetite by magnetotactic bacteria can be mapped onto the generic three-step process discussed in chapter 6 (figure 8.3).

(1) **The sequestering of the reagents required for biomineralisation, and concentration of these.**

Iron ions are sequestered from the environment and taken up by the cell (figure 8.3I). Some researchers have proposed that siderophores are used to aid this process. It is indeed observed that the bacteria can scavenge iron, concentrating large quantities of iron even when it is very scarce in the environment. The most important element for concentrating the iron ions is compartmentalisation. In the magnetotactic bacteria the whole biomineralisation process is carried out within the magnetosome bio–nanoreactor. Lipid analysis has shown high similarities in composition between the magnetosome membrane and the inner cell membrane, indicating that the

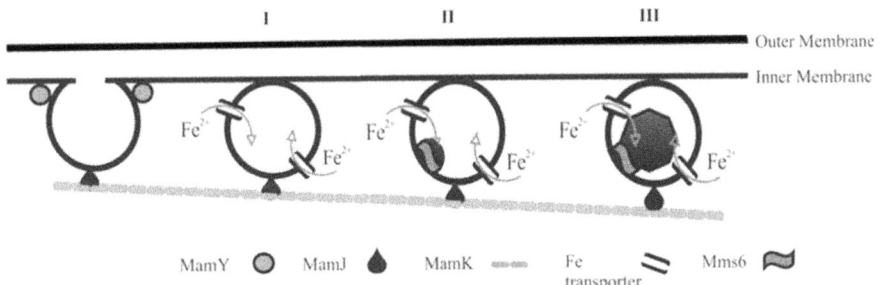

Figure 8.3. Schematic showing the magnetosome formation (adapted from diagram in [35], copyright 2014 with permission from Elsevier) aligned with the generic biomineralisation process. (I) Requesting of ferrous ions; (II) nucleation of magnetite; (III) controlling specific morphological crystal growth.

former is derived directly from the cytoplasmic membrane via a 'budding' process [10]. MamY is the membrane deformation protein (similar to the dynamin and BAR families) which acts to constrict the membrane to initiate vesicle formation [36]. Another protein, Mms16, a GTPase, is thought to hydrolyse GTP to help to form a vesicle-like structure [37]. It is thought that this change in the curvature of the membrane acts to recruit the biomineralisation magnetosome specific protein to the membrane at this point [36], with protein sorting aided by proteins such as MamB. This proposed process is supported by a cryo electron tomographic study, showing that the magnetosomes are attached to the inner membrane, although whether this is the case once the particle has fully matured is not clear [38]. MamK is present as a filament, spanning the long axis of the cell which holds the magnetosomes in the long chains. MamJ anchors the magnetosomes in place by interacting with the MamK fibre. The chain of liposomes serves initially to concentrate the iron ions by importing them across the membrane using iron transport proteins. It is thought the iron is uptaken at all these stages as ferrous ions (Fe^{2+}). The sequestering of the reagents required for biomineralisation, and concentration of these. Specialised iron transporters are capable of transporting iron into the magnetosome against the concentration gradient, from a region of low iron concentration in the cytoplasm to a high concentration within the magnetosome interior. MamB and MamM are found within the magnetosome membrane and are members of the cation diffusion facilitator (CDF) family of membrane transporter proteins, which likely transport iron ions into the magnetosome interior to provide the precursor for magnetite formation [39]. These proteins are antiporters, that transport metal irons in one direction and protons in the other. The formation of magnetite produces protons as a by-product, so these protein drive the precipitation of magnetite not only be providing iron ions, but also be retaining the higher pH required to precipitate magnetite by driving protons out. Magnetotactic bacteria also contain members of the FeoB family of membrane transporters, which are known to transport ferrous iron across membranes [40, 41]. Research which achieved magnetosome formation in a non-magnetic host bacteria, by expressing only the key magnetosome gene, found the magnetosomes formed to be much better quality, larger magnetite particles when the genes encoded FeoB protein were added, showing the importance of good transmembrane iron transport [32].

(2) **The nucleation of the specific inorganic mineral.**

Once a critical iron concentration is reached, chemistry must occur to change the oxidation states of some of the iron ions (convert ferrous (Fe^{2+}) to ferric (Fe^{3+}) ions), to enable magnetite to form (figure 8.3II). A variety of proteins are present in the magnetosome during nucleation and crystal growth, which include MamE, P, and T [42], which are c-like cytochrome redox proteins. These were all also found to be critical in the mamAB knock-out mutagenesis study, suggesting they have a crucial role in controlling the chemistry to allow magnetite to precipitate. Once the chemical conditions are correct, proteins such as Mms6 begin to nucleate the iron into magnetite. Iron continues to be loaded into the magnetosome, allowing the crystal to grow and mature into the correct mineral and with species-specific morphology.

Magnetite nucleation proteins Mms5, 6, 7 and 13.

The proteomic experiment investigating the Mms/Mam protein within the magnetosome membrane found four proteins remain strongly attached to the magnetite crystal, which were named Mms5, Mms6, Mms7, and Mms13, based on their apparent molecular weight (figure 8.2) [34]. All four of these proteins have some common features within their amino acid sequences. All are predicted to be integrated into the membrane due to the regions of their sequences, which are hydrophobic. Mms5, 6, and 7 have one of these regions, whereas Mms13 has two, so is predicted to span the membrane twice, exposing a loop into the centre of the magnetosome to interact with the forming magnetite. Mms5, 6, and 7 all have an $(LG)_n$ amino acid repeating sequence. This leads to a predicted structure that is able to interlock with itself, to self-assemble. Similar repeating sequences have been observed in self-assembly structural scaffold proteins, such as silk protein fibres [43]. Three have an overall isoelectric point (pI) of >8, with the exception of Mms6, which has a pI of approximately 4, giving it a net negative charge at neutral pH. All have a negatively charged (several acidic amino acid residues) region with acidic side chains [34], predicted to be exposed into the centre of the magnetosome to interact with the forming magnetite. For Mms5, 6, and 7 this is the C-terminus end, while it is the central loop in Mms13. Mms6 offers the most acidic region. Several gene knock-out and mutational studies have highlighted the importance of the gene encoding these proteins for the correct formation of the magnetite nanocrystal [44]. The fact that these proteins were tightly bound to the magnetite crystal, combined with what we have learned in chapter 6, concerning proteins offering large arrays of acidic amino acids (in this case through self-assembly), suggest these are nucleation proteins.

(3) **Controlling the crystal growth, specifically with respect to the morphology of the final mineral.**

One of the most remarkable aspects of magnetosome formation is that the morphology is so strictly adhered to (figure 8.3III) within each bacterial strain, but there is so much variety in crystal morphology between strains (see figure 8.1(II)). Unfortunately, however, this is the area of least understanding in the formation process of magnetosomes. The main reason for this is that most of the strains that are culturable and thus readily assessable to work with and research are from the same family (the *Magnetospirillum* family), and as such biomineralise the same cubo-octahedral shaped nanocrystals. The more exotic shaped particles are biomineralised by bacteria that cannot be readily isolated from the environment, let along grown well in a lab or manipulated. The gene sequence and therefore information about predicted proteins is not available. With the advent of extensive rapid and powerful sequencing techniques, there are now some genomes of more elusive magnetotactic bacterial strains with differing particle morphology. However, due to the fact that there is little comparative information to go on, with respect to what sort of sequence patterns we are looking for, it is almost impossible to identify a shape controlling protein from analysis of the genomic sequence alone. So far it has been proposed that MamX is an important protein for controlling crystal growth [45]. A *mms7* knock-out mutation experiment suggests that Mms7 is involved in controlling the crystal morphology [46]. Another protein that is proposed to control crystal morphology is MmsF. MmsF has been described as the master morphological regulator for magnetite biomineralisation *in vivo* [47]. The MmsF protein is encoded by a gene in the same gene cluster as *mms6*. A *mmsF* knock-out mutant has magnetosomes with much smaller, misshapen nanoparticles compared to the control cells. When the *mms6* gene cluster (including the *mmsF* gene) is deleted then a similar but only slightly more severe phenotype is observed. By re-introducing the *mmsF* gene alone back into the $\Delta mms6_{cl}$ strain, the particle morphology is rescued and a near normal magnetosome is observed (figure 8.4) [47]. These experiments indicate a critical role for MmsF in controlling the growth, size and shape of the formed magnetite crystals. MmsF is predicted to have three transmembrane regions, so will display a loop into the interior of the magnetosome similar to Mms13. This loop sequence has a dense region of amino acids with acidic side-chains, but also has another region of more polar and bulky hydrophobic sidechains.

8.4 Biokleptic use of Mms proteins for magnetite synthesis *in vitro*

Proteins within a biological organism are often aided by the other surrounding proteins and the environment in which they exist. However, taking a purified protein outside of this natural environment and studying how it functions *in vitro* (outside the biological environment) is a very powerful technique for understanding how the protein specifically works. Ultimately, this simplifies the system by removing all the other cell machinery, uncoupling the protein function from the rest of the system. This technique is used across the full range of biology to glean crucial information, from enzyme turn over to drug receptor binding. Further, using protein outside the cell within synthetic set-ups can lead to the production of sustainable catalysis and important biotechnological process. In general, a protein can be produced by using molecular cloning techniques to generate larger quantities of protein. This is achieved by inserting the DNA of the protein required into an *E. coli* bacterium to produce (or over express) the protein for us. The protein can then be extracted and purified from the *E. coli.* for *in vitro* use and analysis.

Magnetosome membrane proteins are not easy to over-express and purify, due their amphiphilic nature (typical of transmembrane proteins) and low charge, leading to insoluble protein aggregates, which means a variety of solubilised

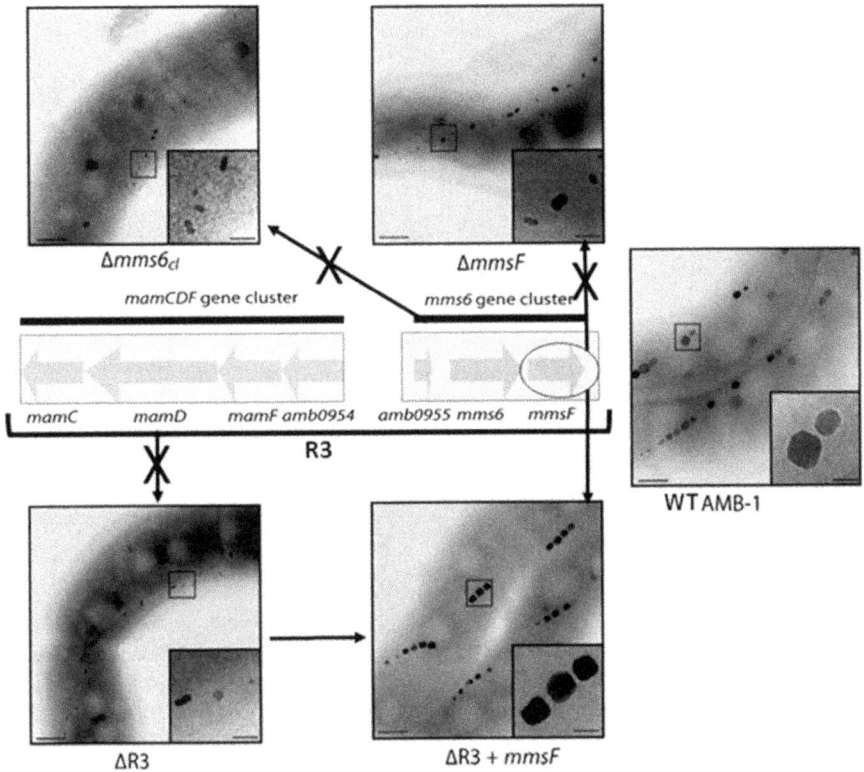

Figure 8.4. Description of the results of the MmsF mutagenesis study. The top left, and right and bottom left TEM images show poor magnetosomes produced, when: the whole mms6 gene cluster is deleted, just mmsF is deleted, and both mamCDFG and mms6 gene clusters (together referred to as R3) are all deleted, respectively. The bottom right image shows the magnetosomes produced in the absence of R3 with the mmsF gene added back in, showing magnetosomes comparable to the wild type (WT) particles (middle right TEM image), showing the MmsF protein alone can result in the activity of the whole R3 region. Scale bars are 100 nm and 50 nm for the inserted images. Adapted from [47], copyright 2012 John Wiley & Sons.

techniques need to be performed. Due to its more acidic nature, Mms6 was the first magnetosome derived protein to be over-expressed in *E. coli* and purified. Mms6 was then added to a room-temperature co-precipitation (RTCP) of magnetite. This reaction is known for its lack of control over the nanoparticles produced, leading to multiple iron-oxide products in a range of sizes and shapes. However, when Mms6 is added to the synthesis, the particles formed are pure magnetite in a regular size of approximately 20 nm with a narrow size distribution [34]. This reaction shows that Mms6 exerts clear control over the particle formation. Analysis has shown the particles are more spherical (similar to cubo octahedral) and also more pure crystalline magnetite compared to control particles [48] (figure 8.5). Furthermore, the ratio of ferrous to ferric ion can be varied, which under normal synthetic conditions directs the formation of alterative iron oxides. However, when Mms6 is added, it can direct the synthesis of magnetite under conditions it would not

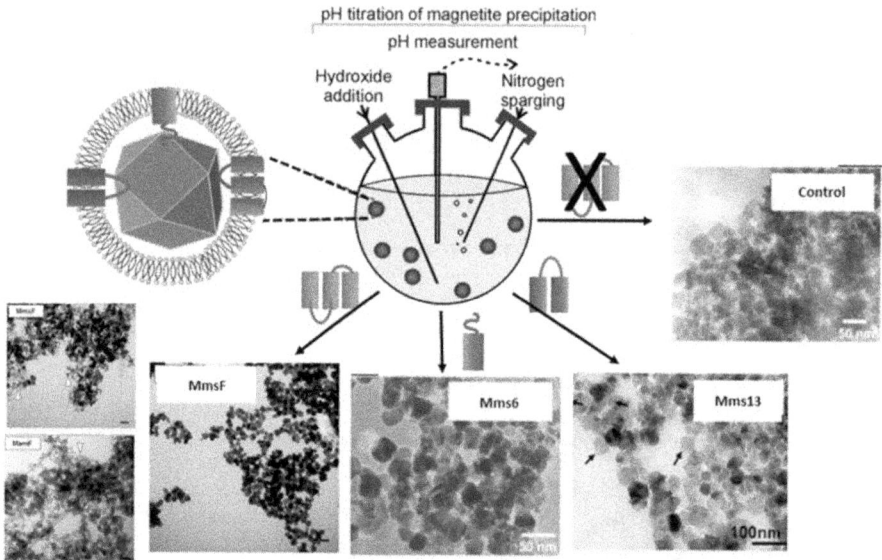

Figure 8.5. Schematic showing how magnetosome membrane (Mms/Mam) proteins can be expressed and purified and added to a chemical co-precipitation of magnetite to control particle formation *in vitro*. Schematic adapted from [49], copyright 2016 John Wiley & Sons, and [52], reprinted by permission from Macmillan Publishers Ltd, copyright 2019. The pink protein represents MmsF, the blue Mms6 and the green Mms13. TEM micrographs show the magnetite particles produced with these protein additives. Scale bars are labelled individually as either 100 or 50 nm. Images adapted from [53], and [54], copyright 2015 with permission of Springer.

normally be formed under in a straightforward chemical synthesis [49]. There are a range of synthetic methods for producing magnetite nanoparticles (see chapter 4). For example, a partial oxidation of ferrous hydroxide (POFHK) produces larger octahedral particles. Interestingly, Mms6 can be added to differing reactions, but Mms6 controls the particles formed to all be approximately 20 nm, regardless of the synthetic method used [48] (table 8.1). Mms6 has also been used in pluronic gel synthesis to slow down diffusion, in an attempt to increase particle size. This was achieved with nanoparticles of around 30 nm being produced, significantly larger than control particles [50, 51].

While more difficult to work with due to its increased hydrophobic nature, Mms13, (also known as MamC), has also been expressed, purified and added to a RTCP reaction (figure 8.5). This also seemed to control the size of the formed particles [54]. Interestingly, the effect was also seen with MamC with a single transmembrane region missing synthesis [54]. When either MamC protein was introduced into a synthetic magnetite precipitation under anaerobic 'free drift' conditions over 30 days, the resulting MNP had an increased average size of 30–40 nm. To date Mms7 has yet to be purified and assessed for activity *in vitro*, but MmsF has been expressed and purified and used in an *in vitro* RTCP reaction for form magnetite. When purified, MmsF was added to synthetic RTCP reactions in amounts consistent with previous Mms6 studies and had a striking effect (figure 8.5). It was found to produce very

Table 8.1. Table to summarize the method and characterisation of Magnetite nanoparticles produced via Mms6 protein mediated synthesis.

	Protein type	MNP synthesis method	MNP size (nm)[a]	Size distribution (nm)[a]	Comments	Reference
Mms6 in solution	Mms6	RTCP (1:1)[b]	20–30 (1–100)	—	Tighter size distribution	[34]
	Mms6	RTCP (1:2)[b]	22.3 (23.1)		Negligible size difference	[55]
	Mms6	POFHK	86 (234)		Smaller mean particle size	[55]
	His$_6$-Mms6	RTCP (1:2)[b]	21.9 (23.1)		Negligible size difference	[55]
	Mms6	POFHK	20 (27.5)	10–30 (10–40)	Smaller mean particle size, Narrower size distribution	[62]
	Mms6	POFHK	20.2 (32.4)	4.0 (9.1)[c]	Narrower size distribution, smaller mean size, cuboidal morphology	[48]
	Mms6	RTCP (1:1)[b]	21.2 (10)	8.3[c]	Larger mean size, cuboidal morphology	[48]
Mms6 in pluronic gels	His$_6$-Mms6	RTCP (2:1)[b]	30	—	Narrow size distribution,	[51]
	His6-Mms6	RTCP (2:1)[b]	36 (4.6)	11 (1.6)	Increased size	[50]
Mms6 peptide in solution	M6A peptide	POFHK	20–25	10–35	Exhibit spherical morphology	[62]
	GLM6A	POFHK	20	10–30		[62]
Surface-bound Mms6	His$_8$-Mms6	POFHK	86 (64)	21(26)[c]	Larger mean particle size, narrower size distribution	[57]
	His$_8$-Mms6	POFHK	90 (69)	15(36)[c]		[58]
	His$_8$-Mms6	POFHK	87 (60)	19 (21)[c]		[56]
Surface-bound Mms6 peptide	Mms6-pep	POFHK	65 (60)	30 (21)[c]	Negligible effect	[56]

(Continued)

Table 8.1. (*Continued*)

	Protein type	MNP synthesis method	MNP size (nm)[a]	Size distribution (nm)[a]	Comments	Reference
MmsF in solution	MmsF-StrepII	RTCP (1:2)	56 (45)	—	Increased size and magnetite purity	[53]
MamC in solution	MamC-His6	Free drift	30–40 (20–30)	20–80	Increased size, and morphology control	[54]
	MamC-truncated-His6	Free drift	30–40 (20–30)	20–80		

RTCP = room temperature co-precipitation, POFHK = partial oxidation of ferrous hydroxide with potassium hydroxide.
[a] Parentheses indicate values for particles synthesised under identical conditions but without protein additive.
[b] Parentheses indicate ratio of ferric to ferrous ions.
[c] Standard deviation. Table from [59].

regular mono-dispersed particles approximately 50 nm in diameter with a very distinctive and defined morphology, suggesting clear crystal growth control [53]. In the same study, two homologues of MmsF (MamF or MmxF (both with highly similar primary and secondary structure to MmsF)) were also added to a reaction and these showed no control over particle morphology. In fact, they produced particles of lower quality with more alternative iron oxides present than the control particles [53] (figure 8.5). This suggests that potentially small residue changes can have significant effects over the type of reaction products which form.

While this section opened with the motive for studying a protein's function through *in vitro* experiment, we have actually seen that this method goes beyond simply assessing the protein's function. We see the protein is actually a biological additive to control a chemical precipitation of magnetite (see sections 7.2 and 7.4 in chapter 7). This is in fact a very attractive middle ground between biological magnetosome biomineralisation and chemical precipitation. In this biomediated method we add key magnetosome synthesis protein to an *in vitro* chemical precipitation of magnetite to control the crystallization. This is a vast improvement on the chemical method with respect to quality. On the other hand, to make magnetosome without the whole bacterial cell and all the cellular machinery that is not relevant (and in many ways opposed) to nanomaterial development is a clear advantage. In this biokleptic synthetic approach nanoparticle yields are vastly increased compared to magnetosomes, and it also offers an increased level of flexibility (removing the cell also removes any issues of toxicity). For example, using a biokleptic Mms6 *in vitro* RTCP synthesis, cobalt can be doped into the particles in a range of increased concentrations, allowed tunable magnetic coercivity [55]. It also offers a new level of control, allowing a choice of which proteins to add to tailor the particles specifically. However, it is worth considering that membrane proteins are difficult to express and purify and thus this biokleptic method with these magnetosome biomineralisation proteins is a long way from being a commercially viable route to manufacture.

8.5 Understanding Mms proteins *in vitro*

It is evident from section 8.8.4 that Mms6, MmsF, and MamC are able to control, to varying degrees, the formation of magnetite MNPs when added to a chemical precipitation. It is clear that an understanding of how these proteins function is now required (as outlined in section 7.2). The majority of the work has been conducted on Mms6 (which is the subject of a review [59]), as this was the first protein to be isolated and used for *in vitro* mediated synthesis and is also the easier (of the options of Mms5, Mms6, Mms7 and Mms13 initially identified) to purify and work with [34]. Interestingly, all four proteins, Mms6, Mms7, MmsF, and MamC, have two distinct commonalities: self-assembly and iron binding capability (figure 8.6).

Iron binding

The first common feature is the presence of multiple acidic amino acids in a clustered region which should be readily accessible to the iron ion precursors and the forming magnetite nanoparticles. All the proteins have these regions with varying degrees of

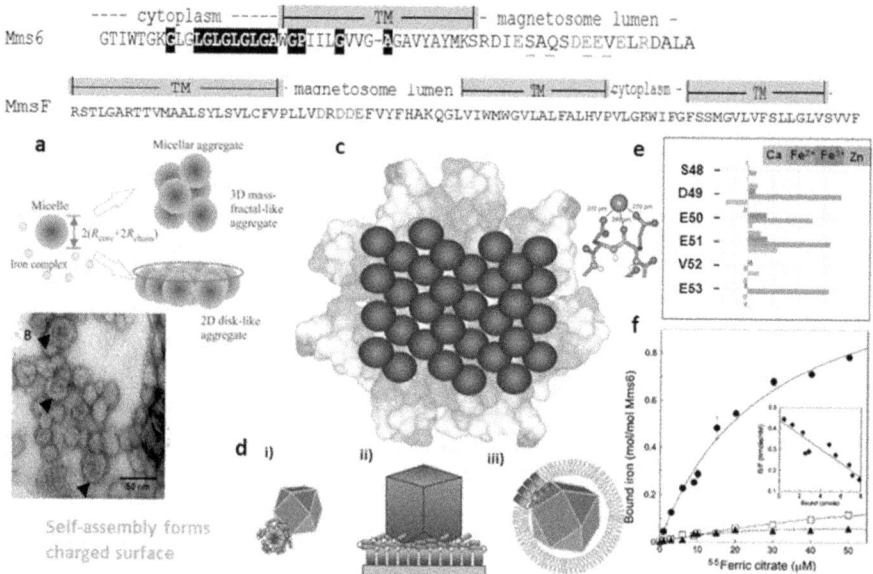

Figure 8.6. Summary of the current understanding of Mms6 and MmsF function. Top: the amino acid sequences for Mms6 and MmsF, annotated with transmembrane region (TM) shown; LG self-assembly repeat section highlighted in black and acidic residues highlighted in red text. (a) and (b) Show the self-assembly feature of Mms6 (micellular) (reprinted with permission from [61], copyright 2015 American Chemical Society) and MmsF (proteinosome) [53] respectively. (c) Bird's eye view schematic of how Mms6 may self-assemble to promote nucleation of magnetite. Right insert shows a computational model of how the acidic residues of Mms6 bind to ferrous ions (reproduced from [49], copyright 2016 John Wiley & Sons). (d) How the curvature of self-assembly may affect particle formation [63], reproduced by permission of The Royal Society of Chemistry. (e) Change in structure of these residues with different metal ions in a 2D NMR [49]. (f) Ferric binding analysis of Mms6 (reprinted with permission [50], copyright 2012 American Chemical Society. Adapted from figure in [59], which in turn uses images from the papers referenced in this caption above.

acidity, with both Mms6 and MmsF having a DEEVE and a DRDDE cluster respectively. Mms6 and MamC (Mms13) have been shown to bind ferric iron ions [34, 43, 50, 54], while Mms6 has been shown to also bind very specifically to ferrous ions [49]. Interestingly, while the looped residue of MmsF can bind iron ions it has a much lower binding affinity than MamC and Mms6 and instead binds much more readily to the magnetite particles.

Self-assembly

All the proteins self-assemble, forming micelles or proteinosomes when purified in aqueous solution. In these structures, the hydrophobic regions of the proteins are believed to be buried within the core of the structure and the hydrophilic, magnetite interaction regions are exposed to the surrounding aqueous environment. When purified these proteins display a secondary structure, suggesting that rather than an amorphous aggregate they actually display a specific folding architecture. This self-assembly is due to the favourable packing of membrane spanning region of the multiple transmembrane region proteins MamC [54] and MmsF [53], and the aggregation through both the amphiphilic nature and the specific repeating LG assembly sequence is responsible for organised self-assembly in the single transmembrane protein Mms6 sequence in the smaller Mms6 and 7. Proteinosome vesicles of MmsF have been observed by TEM [53], while micelles of Mms6 have been visualized [60] and measured with SAXS [61]. Furthermore, the requirement for self-assembly has been tested in a number of experiments. The negative iron interacting region of the protein was used as an isolated peptide, to see if this alone could control particle formation. The peptide without the rest of the protein offered negligible control over particle formation [62]. However, when the iron binding end fused directly to the LG repeating sequence a degree of control was observed, showing how critical the glycine–leucine repeat sections are for self-assembly and thus protein function [62]. In a further similar study, the peptide and the full protein were displayed on a surface to assess if the assembly needed to be controlled or not. In this case again the iron interacting peptide sequence, now randomly assembled on a surface, could not direct mineralisation, but the full-length protein on a surface could [63].

Now we can consider both features of iron ion binding and self-assembly, and extrapolate these findings to inform suggestions about the functions of all these proteins (figure 8.6). For Mms6 and MamC it appears the self-assembly ensures the creation of a proposed iron/magnetite binding surface, which will bind the ferrous and ferric ion to nucleate the formation of magnetite specifically. Interestingly, when particles synthesis is mediated by Mms6 on a surface, the particles formed are much larger. This seems to fit with the hypothesis that Mms6 is a nucleation protein, where the curvature of the surface the magnetite nucleates on directs the size of the particles formed. Indeed, this could explain why the particles formed in solution are so much smaller than magnetosome crystals, as the micelle structure has a very different curvature to the interior of the magnetosome [63]. While MmsF has many of the same features, it is interesting to see that it binds better to the forming crystals and the shape seems to be better controlled with MmsF as an additive. It has been shown that MmsF requires its magnetite interacting region to be constrained in a looped conformation, held in place by two anchored transmembrane sequences. The conformationally free peptide cannot control magnetite synthesis, so we know that structure is critical to the function of MmsF; quite the reverse to Mms6, where the iron interacting/binding end is known to be unstructured. These differences point to MmsF having more of a crystal growth functionality.

8.6 Development and design of additives: emergence of bioinspired magnetite nanoparticle synthesis

8.6.1 Development from biomineralisation proteins: MmsF

We have learnt that MmsF requires a constrained structure, but we also know that Mms proteins are not simple to obtain, and their production is not scalable. This is an excellent example of where we can take the understanding we have to develop to create a bioinspired additive for magnetite synthesis that is much easier to work with

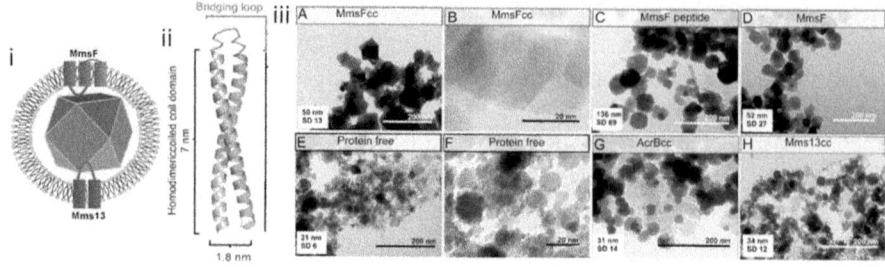

Figure 8.7. (i) Schematic of how MmsF and Mms13 embed in the magnetosome membrane and interact with the magnetite through the loop region. (ii) Structure of the coiled coil (CC) scaffold with the bridging loop region. (iii) TEM images of the particles produced using the CC additive and controls. Adapted from [52] by permission of Macmillan Publishers Ltd.

(section 7.4). Here we know the loop region is required, but the rest of the protein is cumbersome to work with. So, the loop region has been displayed in a constrained loop on a simple 'scaffold' coiled coil protein [52]. This leads to a much more robust, soluble additive that is easier to produce and also more amenable in aqueous chemical reaction conditions. The ease of production and economic benefit of this more soluble additive puts in on the right course for scale-up in the future (figure 8.7).

8.6.2 Screening non-biomineralisation proteins: magnetite interacting proteins

Biopanning (introduced in chapter 5) is a powerful screening technique to discover sequences that bind specifically to minerals (see section 7.2.2). It can be employed to find magnetite specific sequences. However, it can also be employed to find sequences that bind to specific crystal planes of the magnetite. The hypothesis is that biomolecular additives that bind specifically to a specific crystal plan will promote that facet to dominate the morphology, directing towards a particular crystal shape (see sections 4.3 and 4.3.1). To this effect, a protein library of adhirons (displayed on phage) were biopanned against cubic (with [100] crystal faces) magnetite nanoparticles. Adhirons are robust artificial proteins ideal for industry, with a two variable nine amino acid loop region exposed. The hits from this screening were named magnetite interacting adhiron (MIA) proteins. Interestingly, the sequences found to bind most strongly to the [100] magnetite surfaces were very high in basic amino acid residues, with a particular preference for lysine [64]. Adhiron proteins containing both a single loop and two loops showed molecular recognition towards magnetite. This is supported by computational modelling of the amino acids' interaction with magnetite [100] surfaces. Excitingly, these proteins can be used as an additive to direct the formation of cubic nanoparticles when added to RTCP reactions—showing the protein holds the ability to exhibit shape control on the growing crystallization of magnetite nanoparticles [64].

8.6.3 Biomimetic magnetosomes

As we saw in chapter 7, the liposome compartment of magnetosomes can be mimicked in many ways. Indeed, artificial magnetosomes were discussed fully in that section. However, after considering all the work in this chapter, we can use the more detailed understanding of the proteins to improve this method. It was found that while we could use liposomes to form artificial magnetosomes, the use of polymersomes may be more adaptable, flexible, robust, and industrially applicable. We also saw that the whole protein ion channels work effectively to concentrate iron ions into the vesicle; these are very difficult to produce, work with and may not be immediately scalable. A simpler and more effected method of importing iron ions into the polymersomes was to use the physical method of electroporation, which uses electricity to temporary open pores in the polymer membrane. A downside of this is that the reagents (iron ions and base) separated by the membrane immediately mix where they meet at the pore, leading to multiple tiny particles within the membrane itself. Learning from the understanding we have gleaned from the biomineralisation process of magnetosomes and the proteins that perform it, it is clear that a nucleation protein such as Mms6 could aid nucleation within the core of the polymersomes. However, we have also recognised that producing Mms6 is not trivial, and is economically and industrially impractical. Thus we need to mimic the system. We can then consider what we have learnt. We know Mms6 functions by displaying an acidic-rich amino acid surface to nucleate on. Thus, one should design a polymersome with carboxylic acid rich groups covering the interior surface. This was achieved with an asymmetric block co-polymer arrangement of (PEG_{113}–$PHPMA_{400}$) on the outside and ($PMPC_{28}$–$PHPMA_{400}$) directed towards this inside. This asymmetric localisation occurs due to size constraints, with the smaller sized block located on the inner surface and the larger on the outer. PMPC is terminated with the carboxylic group to provide the nucleation surface. When this was electroporated in an iron solution more iron was uptaken and nucleated in the core of the vesicle and thus looks more like a magnetosome than in previous work with non-carboxylated polymersomes [65] (figure 8.8 and both shown for comparison in figure 7.9E).

8.7 Summary: key learning, and the future (towards manufacture)

In this case study we have learnt that there are many ways in which to integrate the biomineralisation of magnetosomes. We can look at the genetics and assess sequences, as well as perform knock-out mutagenesis to investigate which proteins are critical for biomineralisation. We can identify key proteins from their location in the magnetosome membrane, or affinity to the magnetite nanoparticle, and assess these for structure, iron binding ability, etc. Understanding is most powerful when we can use all these techniques to build up a detailed picture of how biomineralisation proteins function to control magnetite nanoparticle growth.

We have also seen how we can use non-biomineralising protein to screen for function that may not be readily seen in Nature. Information extracted from this process can also be used to look for similar sequences in Nature.

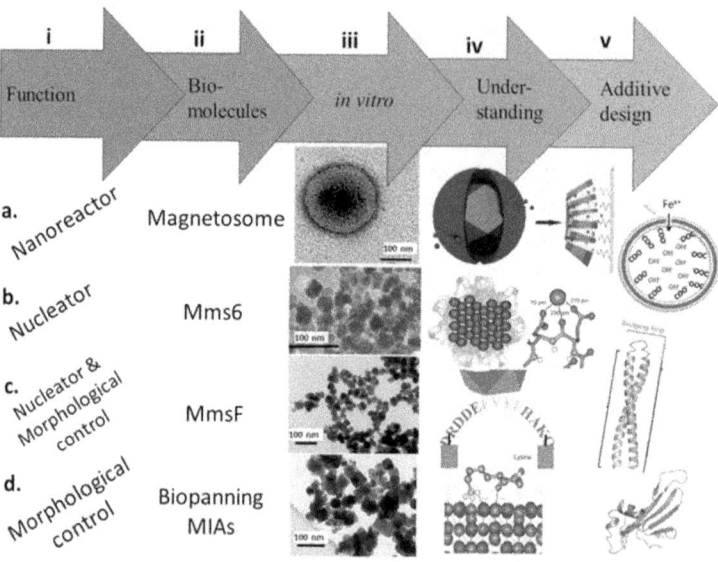

Figure 8.8. Summary figure for this case study. (i)–(v) Show the pipeline so far for magnetite nanoparticles from magnetosomes. Under these headers, an example of all three generic biomineralisation steps are seen. (a) The compartmentalisation is performed by the magnetosome vesicles. Liposome and polymersome artificial examples have been produced. (aiii) A magnetopolymersome *in vitro* that was designed by understanding COOH nucleation sites on Mms6, incorporated COOH sites in the interior (av) and used electroporation (aiv) to import iron ions [65], reproduced by permission of The Royal Society of Chemistry. (b) Nucleation is performed by Mms6 which has been used *in vitro,* and a body or work now describes and understands its function [49], copyright 2016 John Wiley & Sons, and [59]. (c) MmsF seems to control both nucleation and particle shape. It is not fully understood how, but this has too been used *in vitro* as a protein [53] and a more robust additive more suited to scale-up has been produced using a coiled coil scaffold (cv) [52], reprinted by permission from Macmillan Publishers Ltd. (d) The morphological controlling robust MIA protein (dv) was discovered using biopanning. This protein controls the formation of cubic nanoparticles (diii), reproduced by permission of The Royal Society of Chemistry. The positive amino acid residues are thought to be responsible for this by strong interaction with the [100] cubic crystal plane of magnetite (div). Images taken from the references in the caption.

From both of these methods, the trends noted in chapter 6 and the implementation shown in chapter 7 are reinforced. We see some very specific rules occurring, from which we can design future additives.

We see nucleation of magnetite needs an array of acidic amino acids to bind to iron ion. We see with Mms6 that this should be self-assembled in a specific array to obtain maximum effect, but the latest work with polymersomes shows that even just providing this carboxylate charged surface has a nucleation effect.

We see that controlling the crystal growth requires a different set of principles. We see with with both MmsF and the MIA that structural conformation of the biomolecule is essential. Both of these need to be constrained in a loop, perhaps to match the crystal surface for high affinity binding. We also see that basic residues dominate the MIA protein for strong [100] plane cubic magnetite interactions which is demonstrated with computational modelling, whereas the understanding around how the MmsF sequence interaction controls morphology is less clear.

Looking to the future, we can use these principles to design new additives that are more commercially viable. We can adopt these principles into molecules that consider other factors such as robustness to mixing, and other factors of scale-up and manufacture that are not considered in relation to biomineralisation in Nature.

References

[1] Bellini S 1963 Su di un particolare comportamento di batteri d'acqua dolce (About a particular behavior of freshwater bacteria) *PhD Thesis* University of Padua
[2] Blakemore R 1975 Magnetotactic bacteria *Science* **190** 377–9
[3] Schüler D 2007 *Microbiology Monographs* ed A Steinbüchel (Berlin: Springer)
[4] Frankel R B, Blakemore R P and Wolfe R S 1979 Magnetite in freshwater magnetotactic bacteria *Science* **203** 1355–6
[5] Farina M, Esquivel D M S and de Barros H G P 1990 Magnetic iron-sulphur crystals from a magnetotactic microorganism *Nature* **343** 256
[6] Frankel R B, Mann S, Sparks N H C, Jannasch H W and Bazylinski D A 1990 *Nature* **343** 258–61
[7] Lefevre C T *et al* 2013 Monophyletic origin of magnetotaxis and the first magnetosomes *Environ. Microbiol.* **15** 2267–74
[8] Gorby Y A, Beveridge T J and Blakemore R P 1988 Characterization of the bacterial magnetosome membrane *J. Bacteriol.* **170** 834–41
[9] Grunberg K *et al* 2004 Biochemical and proteomic analysis of the magnetosome membrane in Magnetospirillum gryphiswaldense *Appl. Environ. Microbiol.* **70** 1040–50
[10] Tanaka M *et al* 2006 Origin of magnetosome membrane: proteomic analysis of magnetosome membrane and comparison with cytoplasmic membrane *Proteomics* **6** 5234–47
[11] Butler R F and Banerjee S K 1975 Theoretical single-domain grain size range in magnetite and titanomagnetite *J. Geophys. Res.* **80** 4049–58
[12] Dunin-Borkowski R E *et al* 1998 Magnetic microstructure of magnetotactic bacteria by electron holography *Science* **282** 1868–70
[13] Frankel R B, Bazylinski D A, Johnson M S and Taylor B L 1997 Magneto-aerotaxis in marine coccoid bacteria *Biophys. J.* **73** 994–1000
[14] Frankel R, Williams T and Bazylinski D 2007 *Magnetoreception and Magnetosomes in Bacteria* ed D Schüler (Berlin: Springer), 1–24
[15] Xie J, Chen K and Chen X 2009 Production, modification and bio-applications of magnetic nanoparticles gestated by magnetotactic bacteria *Nano Res.* **2** 261–78
[16] Ceyhan B, Alhorn P, Lang C, Schuler D and Niemeyer C M 2006 Semisynthetic biogenic magnetosome nanoparticles for the detection of proteins and nucleic acids *Small* **2** 1251–5
[17] Maeda Y *et al* 2008 Noncovalent immobilization of streptavidin on *in vitro*- and *in vivo*-biotinylated bacterial magnetic particles *Appl. Environ. Microbiol.* **74** 5139–45
[18] Sun J B *et al* 2007 *In vitro* and *in vivo* antitumor effects of doxorubicin loaded with bacterial magnetosomes (DBMs) on H22 cells: The magnetic bio-nanoparticles as drug carriers *Cancer Lett.* **258** 109–17
[19] Hu L L *et al* 2010 Comparison of the $_1$H NMR relaxation enhancement produced by bacterial magnetosomes and synthetic iron oxide nanoparticles for potential use as MR molecular probes *IEEE Trans. Appl. Supercond.* **20** 822–5

[20] Alphandéry E et al 2011 Heat production by bacterial magnetosomes exposed to an oscillating magnetic field *J. Phys. Chem.* C **115** 18–22

[21] Hergt R et al 2005 Magnetic properties of bacterial magnetosomes as potential diagnostic and therapeutic tools *J. Magn. Magn. Mater.* **293** 80–6

[22] Hergt R, Dutz S, Müller R and Zeisberger M 2006 Magnetic particle hyperthermia: nanoparticle magnetism and materials development for cancer therapy *J. Phys. Condens. Matter* **18** S2919–34

[23] Plan Sangnier A et al 2018 Targeted thermal therapy with genetically engineered magnetite magnetosomes@RGD: Photothermia is far more efficient than magnetic hyperthermia *J. Control. Release* **279** 271–81

[24] Staniland S S 2009 *Magnetic Nanomaterials* vol 4 ed C Kumar (New York: Wiley), ch 11 p 399

[25] Heyen U and Schuler D 2003 Growth and magnetosome formation by microaerophilic Magnetospirillum strains in an oxygen-controlled fermentor *Appl. Microbiol. Biotechnol.* **61** 536–44

[26] Sun J B et al 2008 High-yield growth and magnetosome formation by Magnetospirillum gryphiswaldense MSR-1 in an oxygen-controlled fermentor supplied solely with air *Appl. Microbiol. Biotechnol.* **79** 389–97

[27] Staniland S et al 2008 Controlled cobalt doping of magnetosomes in vivo *Nat. Nanotechnol.* **3** 158–62

[28] Tanaka M et al 2012 Highest levels of Cu, Mn and Co doped into nanomagnetic magnetosomes through optimized biomineralisation *J. Mater. Chem.* **22** 11919–21

[29] Alphandéry E, Carvallo C, Menguy N and Chebbi I 2011 Chains of cobalt doped magnetosomes extracted from AMB-1 magnetotactic bacteria for application in alternative magnetic field cancer therapy *J. Phys. Chem.* C **115** 11920–4

[30] Losse A et al 2011 Functional analysis of the magnetosome island in Magnetospirillum gryphiswaldense: The mamAB operon is sufficient for magnetite biomineralization *PLoS One* **6**

[31] Murat D, Quinlan A, Vali H and Komeili A 2010 Comprehensive genetic dissection of the magnetosome gene island reveals the step-wise assembly of a prokaryotic organelle *Proc. Natl. Acad. Sci. U.S.A.* **107** 5593–8

[32] Kolinko I et al 2014 Biosynthesis of magnetic nanostructures in a foreign organism by transfer of bacterial magnetosome gene clusters *Nat. Nanotechnol.* **9** 193–7

[33] Staniland S 2014 An accommodating host *Nat. Nanotechnol.* **9** 163

[34] Arakaki A, Webb J and Matsunaga T 2003 A novel protein tightly bound to bacterial magnetic particles in Magnetospirillum magneticum strain AMB-1 *J. Biol. Chem.* **278** 8745–50

[35] Staniland S S et al 2014 *Nanomagnetism: Fundamentals and Applications* ed C Binns (Amsterdam: Elsevier), ch 3, pp 85–129

[36] Tanaka M, Arakaki A and Matsunaga T 2010 Identification and functional characterization of liposome tubulation protein from magnetotactic bacteria *Mol. Microbiol.* **76** 480–8

[37] Okamura Y, Takeyama H and Matsunaga T 2001 A magnetosome-specific GTPase from the magnetic bacterium Magnetospirillum magneticum AMB-1 *J. Biol. Chem.* **276** 48183–8

[38] Komeili A, Li Z, Newman D K and Jensen G J 2006 Magnetosomes are cell membrane invaginations organized by the actin-like protein MamK *Science* **311** 242–5

[39] Uebe R *et al* 2011 The cation diffusion facilitator proteins MamB and MamM of Magnetospirillum gryphiswaldense have distinct and complex functions, and are involved in magnetite biomineralization and magnetosome membrane assembly *Mol. Microbiol.* **82** 818–35

[40] Rong C *et al* 2008 Ferrous iron transport protein B gene (feoB1) plays an accessory role in magnetosome formation in Magnetospirillum gryphiswaldense strain MSR-1 *Res. Microbiol.* **159** 530–6

[41] Rong C *et al* 2012 FeoB2 Functions in magnetosome formation and oxidative stress protection in Magnetospirillum gryphiswaldense strain MSR-1 *J. Bacteriol.* **194** 3972–6

[42] Siponen M I, Adryanczyk G, Ginet N, Arnoux P and Pignol D 2012 Magnetochrome: a c-type cytochrome domain specific to magnetotactic bacteria *Biochem. Soc. Trans.* **40** 1319–23

[43] Feng S *et al* 2013 Integrated self-assembly of the Mms6 magnetosome protein to form an iron-responsive structure *Int. J. Mol. Sci.* **14** 14594–606

[44] Tanaka M, Mazuyama E, Arakaki A and Matsunaga T 2011 MMS6 protein regulates crystal morphology during nano-sized magnetite biomineralization *in vivo J. Biol. Chem.* **286** 6386–92

[45] Yang J *et al* 2013 MamX encoded by the mamXY operon is involved in control of magnetosome maturation in Magnetospirillum gryphiswaldense MSR-1 *BMC Microbiol.* **13** 203

[46] Yamagishi A *et al* 2016 Control of magnetite nanocrystal morphology in magnetotactic bacteria by regulation of mms7 gene expression *Sci. Rep.* **6** 29785

[47] Murat D *et al* 2012 The magnetosome membrane protein, MmsF, is a major regulator of magnetite biomineralization in Magnetospirillum magneticum AMB-1 *Mol. Microbiol.* **85** 684–99

[48] Amemiya Y, Arakaki A, Staniland S S, Tanaka T and Matsunaga T 2007 Controlled formation of magnetite crystal by partial oxidation of ferrous hydroxide in the presence of recombinant magnetotactic bacterial protein Mms6 *Biomaterials* **28** 5381–9

[49] Rawlings A E *et al* 2016 Ferrous iron key to Mms6 magnetite biomineralisation: A mechanistic study to understand magnetite formation using pH titration and NMR *Chem. Eur. J.* **22** 7885–94

[50] Wang L J *et al* 2012 Self-assembly and biphasic iron-binding characteristics of Mms6, a bacterial protein that promotes the formation of superparamagnetic magnetite nanoparticles of uniform size and shape *Biomacromolecules* **13** 98–105

[51] Prozorov T *et al* 2007 Protein-mediated synthesis of uniform superparamagnetic magnetite nanocrystals *Adv. Funct. Mater.* **17** 951–7

[52] Rawlings A E *et al* 2019 Artificial coiled coil biomineralisation protein for the synthesis of magnetic nanoparticles *Nat. Commun.* **10** 2873

[53] Rawlings A E *et al* 2014 Self-assembled MmsF proteinosomes control magnetite nanoparticle formation *in vitro Proc. Natl. Acad. Sci. U.S.A.* **111** 16094–9

[54] Valverde-Tercedor C *et al* 2015 Size control of *in vitro* synthesized magnetite crystals by the MamC protein of Magnetococcus marinus strain MC-1 *Appl. Microbiol. Biotechnol.* **99** 5109–21

[55] Galloway J M *et al* 2011 Magnetic bacterial protein Mms6 controls morphology, crystallinity and magnetism of cobalt-doped magnetite nanoparticles *in vitro J. Mater. Chem.* **21** 15244–54

[56] Bird S M, Rawlings A E, Galloway J M and Staniland S S 2016 Using a biomimetic membrane surface experiment to investigate the activity of the magnetite biomineralisation protein Mms6 *RSC Adv.* **6** 7356–63

[57] Bird S M, El-Zubir O, Rawlings A E, Leggett G J and Staniland S S 2016a A novel design strategy for nanoparticles on nanopatterns: interferometric lithographic patterning of Mms6 biotemplated magnetic nanoparticles J. Mater. Chem. C **4** 3948–55

[58] Bird S M, Galloway J M, Rawlings A E, Bramble J P and Staniland S S 2015 Taking a hard line with biotemplating: cobalt-doped magnetite magnetic nanoparticle arrays *Nanoscale* **7** 7340–51

[59] Staniland S S and Rawlings A E 2016 Crystallizing the function of the magnetosome membrane mineralization protein Mms6 *Biochem. Soc. Trans.* **44** 883–90

[60] Kashyap S, Woehl T J, Liu X, Mallapragada S K and Prozorov T 2014 Nucleation of iron oxide nanoparticles mediated by Mms6 protein *in situ ACS Nano* **8** 9097–106

[61] Zhang H *et al* 2015 Morphological transformations in the magnetite biomineralizing protein Mms6 in iron solutions: a small-angle x-ray scattering study *Langmuir* **31** 2818–25

[62] Arakaki A, Masuda F, Amemiya Y, Tanaka T and Matsunaga T 2010 Control of the morphology and size of magnetite particles with peptides mimicking the Mms6 protein from magnetotactic bacteria *J. Colloid Interface Sci.* **343** 65–70

[63] Bird S M, Rawlings A E, Galloway J M and Staniland S S 2016 Using a biomimetic membrane surface experiment to investigate the activity of the magnetite biomineralisation protein Mms6 *RSC Adv.* **6** 7356–63

[64] Rawlings A E *et al* 2015 Phage display selected magnetite interacting Adhirons for shape controlled nanoparticle synthesis *Chem. Sci.* **6** 5586–94

[65] Bain J *et al* 2019 A biomimetic magnetosome: formation of iron oxide within carboxylic acid terminated polymersomes *Nanoscale* **11** 11617–25

Chapter 9

Case study 2: silica

9.1 Biosilica occurrence and formation

Biological silica has interested scientists for centuries, due to the intriguing structures of biosilica formed, as well as the nature of the entire process of their formation. There have been numerous studies on finding, documenting, and studying ornate biosilica structures. Of particular interest are diatoms, due to their hierarchically-structured silica cell walls, and the variety of structures seen. In fact, in the early days of their study, diatoms were classified based on their shapes, morphologies and structures (see figure 9.1). Biosilica is also found in radiolaria, sponges and many plants. Diatoms and radiolaria form silica-based cell walls, which provide protection and are porous in order to allow transport of nutrients and signals. Sponges use needle-like spicules that offer mechanical support to their bodies, in addition to providing protection against predators. Plants, on the other hand, can produce silica as a by-product of the uptake of silicon from soil. However, this *side reaction* has been used to plants' advantage in many cases. One example is rice, where the leaves become strong due to the presence of silica and hence maximise the absorption of sunlight (this is described in section 6.2.3). Other plants, e.g. cucumber, use biosilica formation in response to a threat or an attack and contain the pathogens in a biosilica *cage*.

In the last 15–20 years, scientists have navigated from morphological observations to molecular and genetic mechanisms. A number of studies on cell biology and molecular biology had made great contributions to our understanding of biological silica formation (figure 9.2). The process of biosilica formation is described briefly here, and readers are referred to the further reading section for references containing detailed descriptions of biosilicification [3–10]. The key stages involved are:
  1. uptake of silicon into the living system,
  2. its transport inside the system,
  3. biochemical transformation into (intermediate and) final form, and
  4. deposition into the final location.

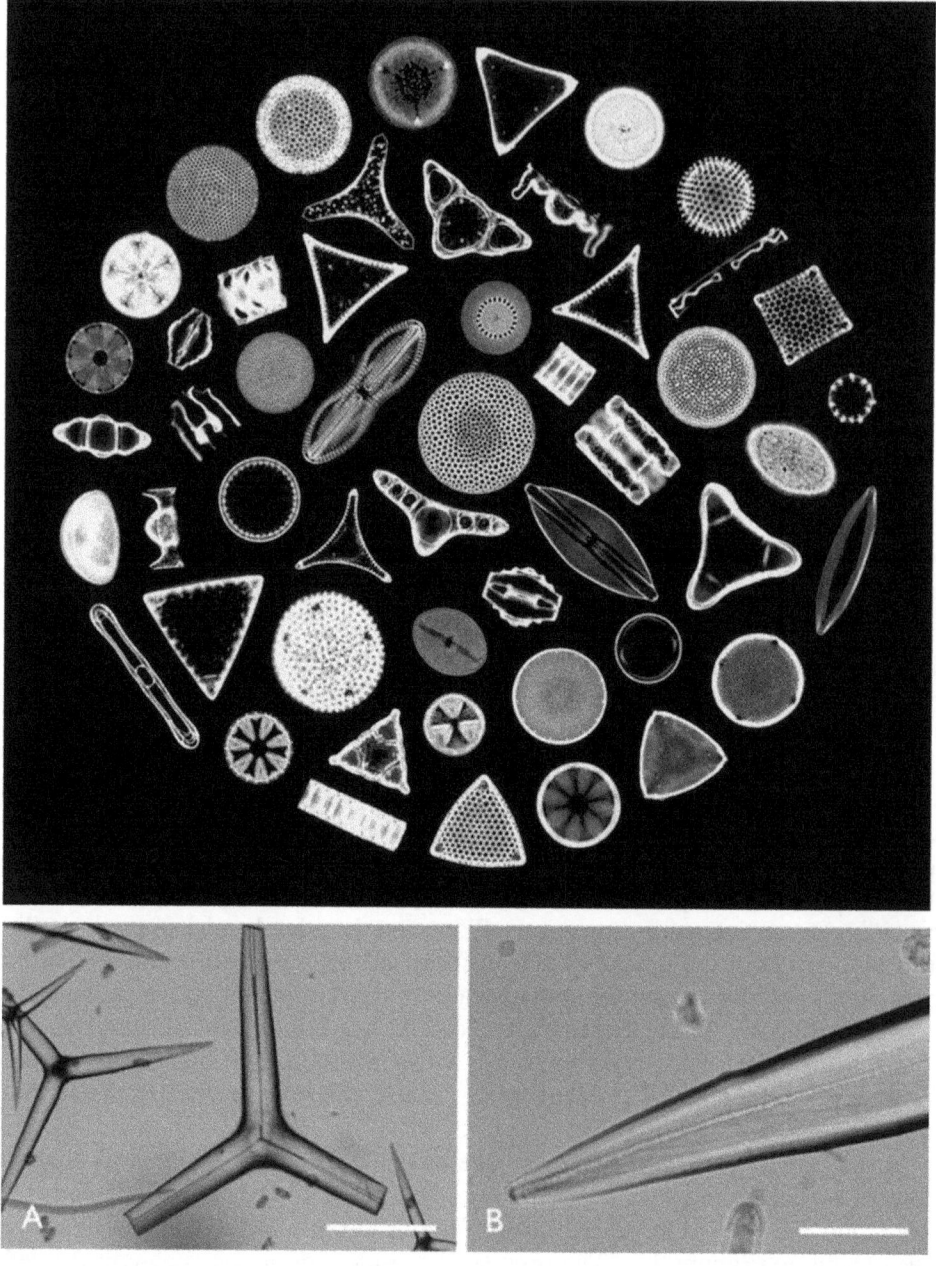

Figure 9.1. Top: morphologies of various diatoms (credit: Randolph Femmer, USGS [1]). Bottom: silica spicules in sponges (image reproduced with permission from [2], copyright 2003 John Wiley & Sons).

This is the general order of the process, however, variations may occur from system to system. For example, it is not fully clear whether the transport within the organisms occurs in the form of orthosilicic acid (a silica monomer), silicic acid oligomers, or as *final* silica. Further, in some examples, steps 3 and 4 can be

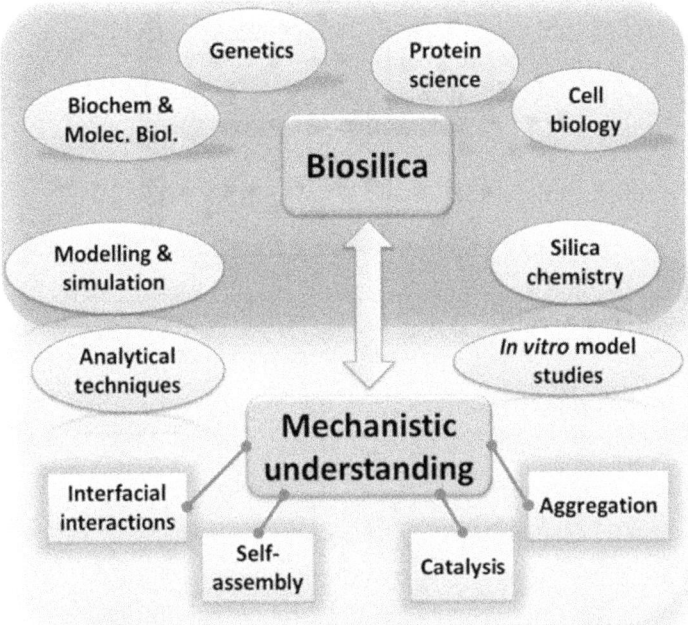

Figure 9.2. Schematic representation of various tools and approaches used to study biosilica and its formation, leading to mechanistic understanding. Image taken from [11], reproduced by permission of The Royal Society of Chemistry.

combined into a single step, where the chemical transformation to silica and deposition occur simultaneously. It is worth noting that each step is under strict biological and genetic control, particularly the order of events and the secretion of the signals/biomolecules governing each step. Given the significant progress made in understanding the processes in diatoms over other biosilica systems, most of the examples below include those from diatoms.

The uptake of silicon in most cases is active, which means that when silicon is needed for growth or cell division, the organism actively and selectively *absorbs* silicon from its environment (figure 9.3). Various organisms have different mechanisms to achieve this, however, they are remarkable in their selectivity. This selectivity allows the organisms to concentrate silicon from a highly unsaturated (~70 μM) extra-cellular concentration to a near-saturated (1–2 mM) or supersaturated intra-cellular concentration (19–340 mM for some diatoms). This is possible due to specific recognition of the soluble monomeric form (silicic acid), as well as the ability to stabilise oligo-/polymeric forms to avoid premature silica deposition. This stabilisation of supersaturated *pools* of silicon allows the organism to effectively supply feedstock to the site of transformation and deposition. This feedstock is then biochemically converted or polymerised into biosilica and deposited in the final form. The final product is typically an organic–inorganic hybrid of biomolecules and silica.

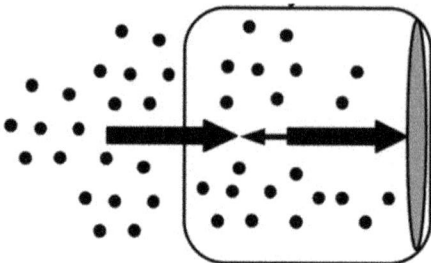

Figure 9.3. Controlled uptake of silicic acid (shown as dots) in diatoms (shown as a box) from outside to inside the cell and to the deposition site (shown as grey elongated oval biosilica). The arrows indicate the direction and quantity (thickness) of transport. Image taken with permission from [8], copyright 2008 American Chemical Society.

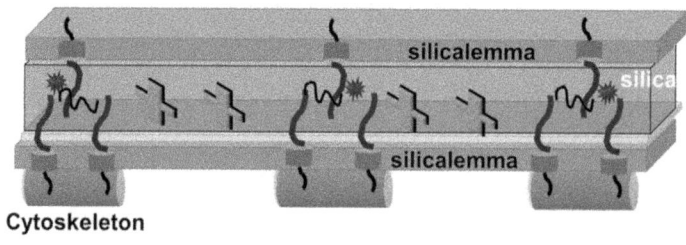

Figure 9.4. A simplified representation showing the proposed localisation of various biomolecules associated with biosilica formation in diatoms. Figure adapted from [40].

9.2 Biomolecules controlling biosilica formation

Each of the stages discussed above involve highly specialised biomolecules—their functions range from recognising monomeric silicon for transport, complexing with intermediate silicate oligomers for stabilisation and catalysing or templating the formation of biosilica from precursors. The molecular/chemical details for specific organisms are discussed in this section. In the late 1990s and early 2000s, the key molecules involved in silicon transport and polymerisation were identified for diatoms, plants and sponges [8, 12–23]. Genomic investigations of diatoms and sponges have provided deeper understanding of biosilicification [19, 24, 25].

A number of biomolecules, including proteins, have been isolated from biosilica of diatoms, sponges and plants. These include, for example, silaffin (a family of silica forming proteins), SiMat proteins, propylamines, silicalemma-associated proteins (SAPs), silacidin, cingulins, and polysaccharides from diatoms [26–32], silicatein and silintaphin proteins from sponges [15, 16, 33–35], and a range of biomolecules from plants [12, 17, 36–38] and choanoflagellates [39]. Through extensive research, the roles of these biomolecules in biosilica formation are becoming clear. Based on the expression, occurrence and function of these organic components, figure 9.4 shows

the proposed localisation of various organic molecules associated with biosilica formation in diatoms [40]. Broadly, the functions of these components can be generalised into the following categories:
- framework or scaffold,
- catalytic, and
- regulatory.

The framework components are generally found anchored within a membrane (e.g. the silicalemma in figure 9.4) or the cytoskeleton. These components provide a rigid support for the catalytic and regulatory components to reside on. The catalytic components (e.g. silaffins and propylamines in diatoms, and silicatein in sponges) carry out the chemical transformation of silicic acid (or other form of silica precursors) to silica particles. The regulatory components (e.g. silacidin proteins) help control the biosilica formation (e.g. the amount of silica particles formed). It is proposed that the amine functionalities such as lysine or arginine side groups and propylamines participate in the catalysis of silica polymerisation. Self-assembly of the organic components and organic–inorganic electrostatic interactions also play a key role in biosilica formation. The framework components also control the location for biosilica formation, while other dynamic processes can help move these freshly formed biosilica particles and enable their deposition to the desired sites. The entire process is very complex, with temporal and spatial control over each step.

It is clear that compared to other systems, research on diatom biosilica formation has progressed the furthest. However, silicatein, an enzyme that catalyses silica formation in sponges, has been isolated and studied. Unlike biosilica formation in diatoms, sponges appear to use very few components for biosilica formation, although the chemical mechanisms are quite different, given that diatom biosilica is not formed using enzymes. Silicatein has a specific catalytic site composed of serine, histidine and asparagine.

It is proposed that the association of silicic acid with the hydroxyl group of serine and the imidazole group of histidine is crucial in the enzymatic catalysis. Although the exact details of this mechanism are not fully understood, there are two hypotheses, see figure 9.5. The first mechanism (figure 9.5, top), proposes that silicatein hydrolyses the precursor, thereby activating it for further (non-enzymatic) autocatalytic reaction. The second mechanism is based on biocatalysis of the condensation of fully hydrolysed monomeric silica (silicic acid, as a substrate) to cyclic trimers, which undergoes further non-enzymatic polymerisation. Despite these differences, both mechanisms rely on chemical *recognition* (which could be covalent bonding) between the silica precursor and –OH of serine, and the imidazole N of histidine of the catalytic site of silicatein.

An important reason for investigating biosilicification is to learn from biology how to design sustainable technologies for novel materials for applications. In order to transfer the knowledge obtained from biosilicification, we have seen in chapter 7 that scientists have taken a three step approach:
  1. Identify, isolate and characterise biomolecules and genetic machinery controlling biosilica formation.

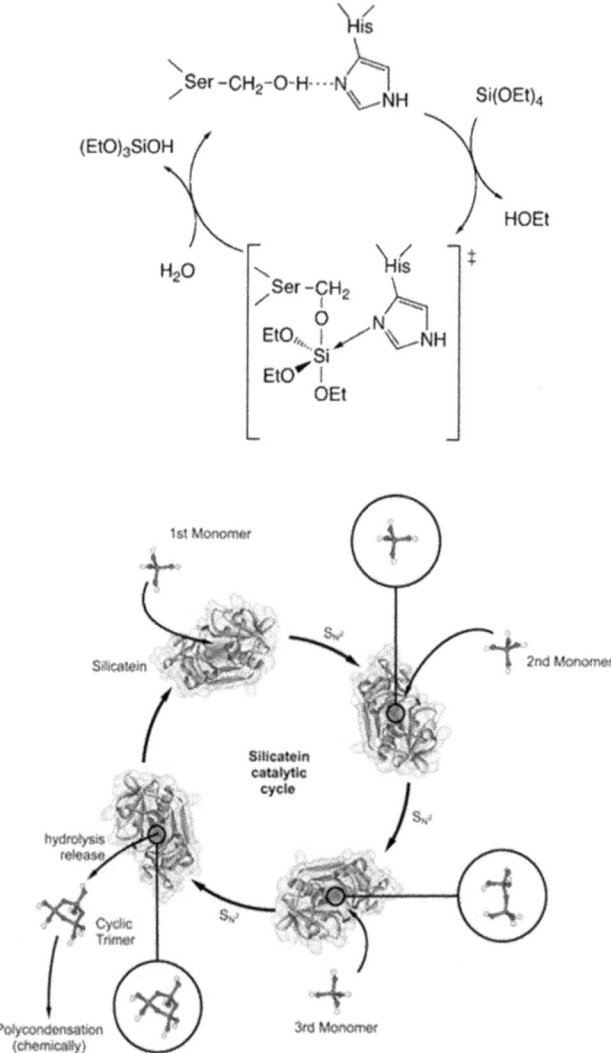

Figure 9.5. Proposed two mechanisms for enzymatic action of silicatein. Images reprinted with permission from [41], copyright 2008, American Chemical Society (top) and [42], copyright 2016, American Chemical Society (bottom).

2. Perform *in vitro* model studies using these bioextracts (and other strong-binding peptides) to understand the roles of various key functionalities.
3. Use synthetic molecules—bioinspired custom-made additives—to mimic the function of bioextracts/peptides and design sustainable processes.

Step 1 is already discussed briefly above, in section 9.2. Below we discuss the next two steps and illustrate them with selected examples.

9.3 Learning from biological silica synthesis: *in vitro* investigation of bioextracts

Once the biomolecules responsible for biosilica formation have been isolated and purified, their roles in the formation can be probed *in vitro*. Such research allows a deeper understanding of the physicochemical properties and function of the bioextracts, and also enables the reconstruction of the biosilicification process (such as that shown in figures 9.4 and 9.5). This can be illustrated with the example of silicateins from sponges.

As shown in figure 9.5, the catalytic triad, responsible for catalysing silica formation, was studied *in vitro*, using a range of conditions and precursors (or substrates). In addition, when the key amino acids from the catalytic site were replaced with other residues, the enzymatic action was lost. Further, the analysis of the crystal structure of the enzyme, molecular modelling, and simulation of the triad and its function provided insights into the catalytic action. It became clear from these studies that the close proximity of serine and asparagine with imidazole N from histidine is important to localise silicic acid via hydrogen bonding, while making it reactive by deprotonating simultaneously. Silintaphin-1 protein is postulated to act as a co-factor to silicatein and enhance the catalytic function. In addition to the chemical or catalytic attributes, silicateins also exhibit unique physical properties—they cooperatively self-assemble with silintaphin-1 into long filaments which form the templates for the hallmark needle-like spicules in sponges. These investigations have clearly identified the key catalytic function of the triad and the importance of self-assembly for templating.

In the case of diatoms, as the extracted biomolecules became available, their roles in silica formation were investigated *in vitro* [29, 43–46]. Initially, these biomolecules were studied individually, however, as additional components were identified and characterised, they were also studied in combination with each other. Similar to the case of sponge silicateins, silaffin (and other) proteins showed unique activities in

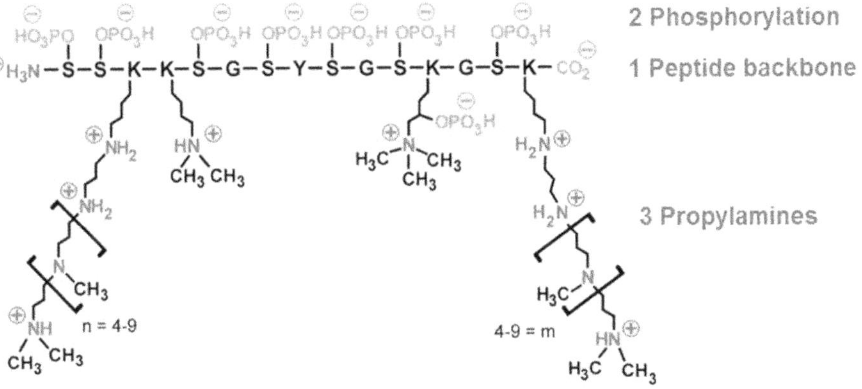

Figure 9.6. Structure of one of the silaffins, showing three key functionalities. Image taken from [29], reprinted with permission from AAAS.

catalysing, regulating, and templating biosilica formation [29, 47]. Scientists deconstructed various key features of those proteins and tested their function individually. An early example of this is the family of silaffin proteins (see figure 9.6), which contain (1) a unique sequence of amino acids in the peptide backbone, and various modifications such as (2) phosphorylation, and (3) propylamine functionalisation.

Each of these three features were studied for their roles in silica formation *in vitro*, using a range of experimental and computational techniques. These included the direct use of the peptide and its mutants (without any modifications) in silica formation, NMR studies of the self-assembly of these peptides (with and without the presence of silicates), and the study of the effects of synthesis conditions (e.g. pH, solvents, counter-ions, and silica precursors) [48]. These studies highlighted various aspects of silaffins that are key to biosilica formation. The amine moieties, the peculiar protonation states and their geometric arrangement lead to a highly cationically charged molecule which creates a favourable micro-environment for silica formation, and also exhibits catalytic action during the condensation reactions. In addition, these biomolecules form self-assembled structures in solution given their specific structure, which template silica formation. The phosphorylation is thought to help anchor the proteins and/or cross-link them with polysaccharides (or other cellular components), in order to exert spatial control over silica formation.

9.4 Emergence of bioinspired synthesis using synthetic 'additives'

The knowledge of the physicochemical properties and functions of the bioextracts have inspired a huge amount of 'synthetic' research on silica formation *in vitro*. These studies can lead to improved understanding of silica (and biosilica) formation in the presence of such additives (and bioextracts). They can further help design sustainable technologies for commercially relevant materials. To achieve these goals, a successful approach taken includes the design and use of synthetic molecules (defined as additives in chapter 7), which mimic the function of these biomolecules. Another approach involves the direct use of bioextracts. The latter provides a great level of control, given that the biomolecules have been optimised via evolution for biosilica formation. However, the issues with their isolation, purification, and stability can limit their use due to significant costs and the limited availability associated with this. To date, none of these biomolecules have been available at quantities of tens or hundreds of grams, while their access is limited to those who can extract them or express recombinantly. Further, it is likely that regulatory approval for their use in manufacturing and consumer products could be problematic. As a result, with industrial use in mind, the former approach of using additives has the potential to be economical, sustainable, scalable, and safe.

Continuing with the examples of silicateins and silaffins from previous sections, let us illustrate how bioinspired silica synthesis was developed, and how the additives were designed. As we have seen in silicateins, both the catalytic residues and the self-assembly of the proteins is essential, and so it could be prudent to design synthetic systems which can offer these features. Moving on to the examples of diatoms, we have seen that certain amino acids (e.g. lysine or arginine) and propylamines provide

a highly cationic nature to silaffins. Learning from this, scientists explored the answers to questions pertaining to *in vitro* silica formation, such as [48–57]:
- Which amino acids are important?
- Are the amino acid sequences critical to silica formation?
- Would (homo)polypeptides be sufficient to promote silica formation?
- Do we need peptides or biomolecules?
- Can smaller molecules provide similar activities?
- Is carbon–carbon spacing important in the propylamines?
- Are the number of amines (or propylamine chain length) and amine protonation important?
- ....

There have been numerous investigations addressing these and other questions. Below, we provide a summary of those reports in an attempt to answer most of the questions listed above.

9.4.1 Which amino acids are important?

Many approaches have been taken to further the knowledge of the roles of biomolecules in silica formation. Some investigations focused on the use of individual amino acids in order to identify those that have the ability to catalyse and/or template silica formation. A novel system was designed to address this, in the context of mimicking the activity of silicateins [58]. Two sets of gold nanoparticles were functionalised with imidazole or hydroxyl groups (see figure 9.7). Although this strategy does not directly mimic the catalytic triad, the enhanced silica formation in the presence of these two nanoparticles highlighted the importance of co-operative action by hydroxyl and imidazole functionalities. In other studies, amino acids were also individually used (unbound to any particles) and their roles in silica formation were studied [52, 57]. It emerged that basic amino acids (lysine, arginine and histidine) were able to catalyse the silicic acid condensation reactions.

9.4.2 Would (homo)polypeptides be sufficient to promote silica formation?

The identification of key amino acids inspired the use of respective polypeptides to study their effects on silica formation. As seen in figure 9.8, polypeptides each of these amino acids (lysine, arginine, and histidine) were able to form silica particles within seconds to minutes at ambient conditions, in water at pH 7. It is worth highlighting that until that point, the synthesis of such particles required either heating, alkaline pH, a co-solvent and/or a much longer time.

Polylysine is a peculiar example which combines the physical and chemical properties of biomolecules. In addition to their catalytic function, leading to spherical silica particles, polylysine was able to produce hexagonal silica plates ~50 nm thick and 100s nm to one micron size sides (figure 9.9, top) [59–62]. Such structures had never been reported to form under such mild conditions. When the polylysine conformation, and its dynamic and co-operative self-assembly with silica species in solution were studied, it was found that through charge–charge

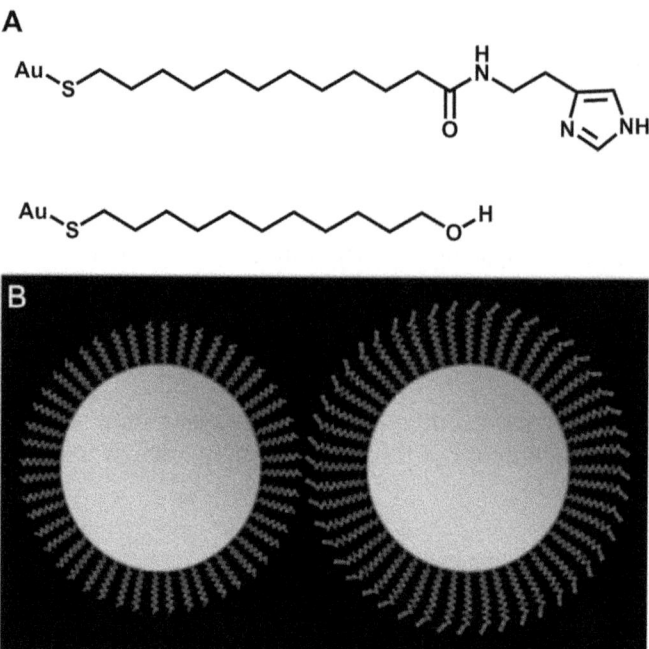

Figure 9.7. The molecules in (a) are used to functionalise two separate sets of gold nanoparticles in (b). Image reproduced with permission from [58], copyright 2005 John Wiley & Sons.

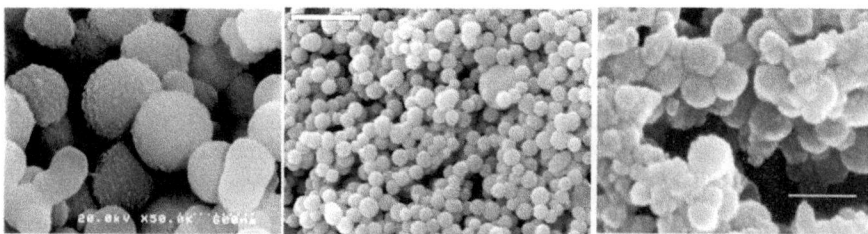

Figure 9.8. SEM images of silica particles formed in the presence of poly-lysine (left), poly-arginine (middle) and poly-histidine. Bars = 600 nm (left), 2 μm (middle) and 300 nm. Images taken from [56] (middle) and [55] (right), copyright 2003 with permission of Springer.

interactions and charge-screening, random coils of polylysine co-assemble with silicate ions into helical conformation (figure 9.9, bottom). A number of polylysine helices pack together into hexagonal single crystals, which template condensing silicates, thus forming hexagonal silica. In another study, a range of block co-polypeptides were designed which contained a catalytic block and a self-assembling block [50]. As expected, these polypeptides provided self-assembled templates for silica formation, which was catalysed by the catalytic blocks. We will expand this strategy later in this chapter. This incredible ability to produce structured materials at ambient conditions is of significant interest for designing high-value materials, as well as for developing responsive materials, such as those that can 'switch' their morphologies [60, 63].

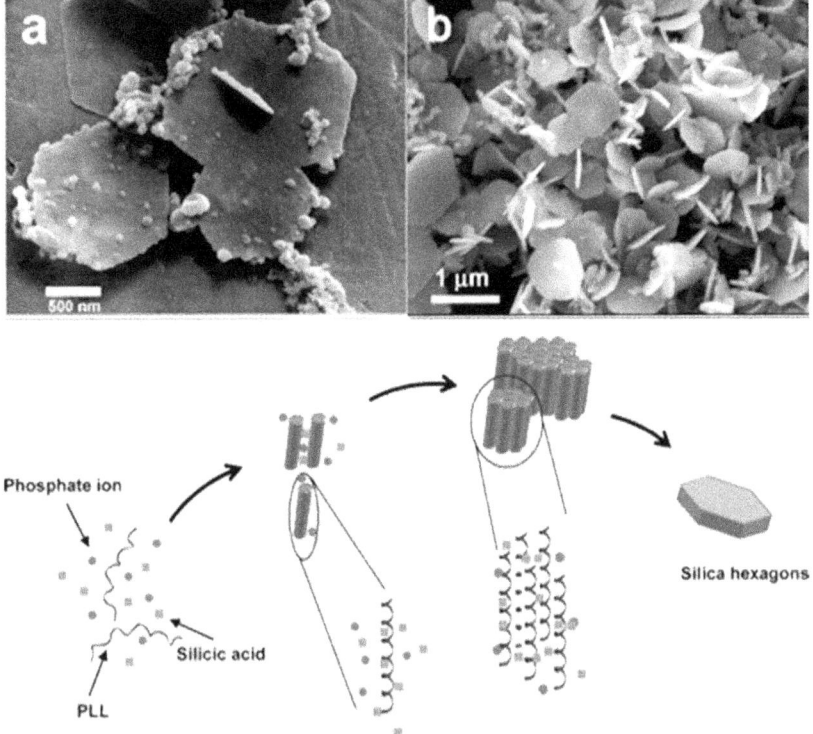

Figure 9.9. Illustration of how additives (polylysine—PLL in this case) can self-assemble into intricate templates and catalyse the formation of interesting bioinspired silica structures. Top: SEM micrographs of hexagonal silica formed using polylysine. Bottom: schematic presentation of the self-assembly, where the blue rods denote helical PLL. Images taken with permission from [60], copyright 2006, (top) and [62], copyright 2005, (bottom), American Chemical Society.

9.4.3 Peptides from biopanning

As discussed in chapter 7 (section 7.2.2 and table 7.1), a number of strong binding peptide sequences were identified against silica surfaces and particles. Although some of these peptides had some (limited) similarity with those isolated from biosilica (e.g. diatoms), there was less than 20% similarity between the peptides reported as strong silica binders. This variation was puzzling, particularly because most other materials, especially metals, have provided consensus in sequences. Therefore, in order to obtain the principles of the peptide–silica interface, further in-depth research was performed [64–67].

In one such study [64], silica particles of three distinct sizes were synthesised using the same method (figure 9.10, top). Using biopanning, strong binding peptides were identified against these particles. Interestingly, there was little similarity between these sequences, as well as with previously reported sequences. Extensive analysis of the surface chemistry using experiments, modelling, and simulation revealed that the particle surfaces have different ionisation. This was attributed to the slight differences in ammonia concentration used during their synthesis as a way to control

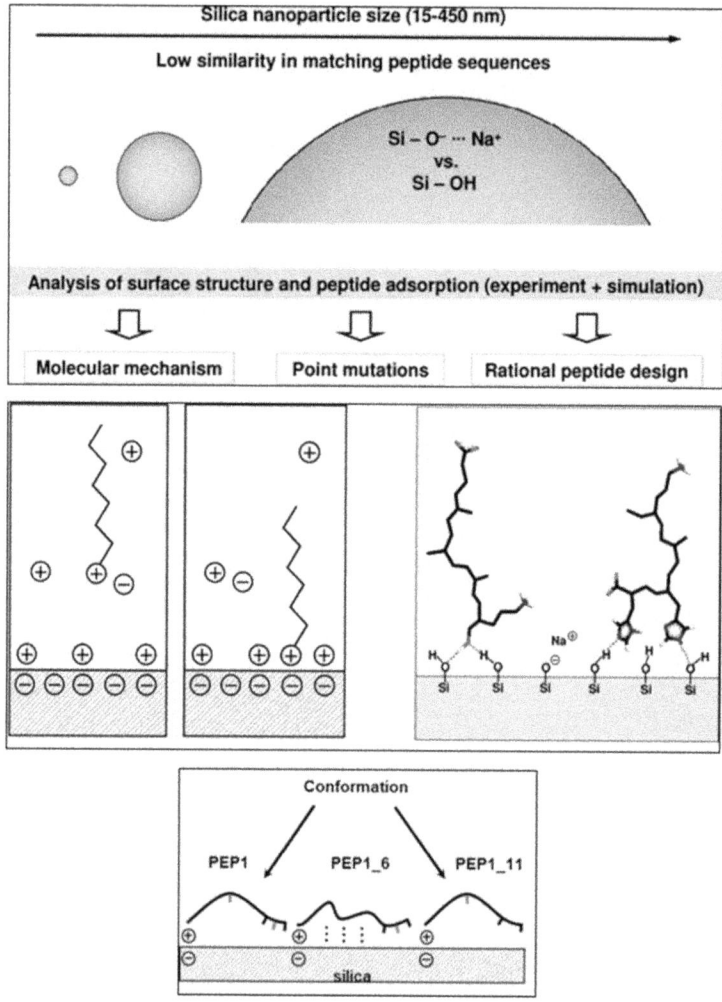

Figure 9.10. Top: an outline of how peptide binding on silica particles of three different sizes was investigated. The middle image shows that the major contributions to peptide adsorption comes from ion pairs (left) and (right) hydrogen bonds. Bottom: replacing a key amino acid (shown in red) leads to the modulation of binding through changes in peptide conformation and available binding sites on the peptide. Images reprinted with permission from [64], copyright 2012 American Chemical Society.

particle sizes. It was striking that such subtle differences were recognised by the peptides. Upon performing a range of peptide binding experiments and simulations, it became clear that the peptides bind via ionic interactions through unoccupied surface charges/sites. Further hydrogen bonds are also formed that provide additional binding strength (figure 9.10, middle). In the next step, the focus was on the role and position of individual amino acids. As we have seen above, histidine has been suggested as a key amino acid in silica binding, while it also consistently appeared in silica-binding peptides. By creating mutant peptides where histidine was replaced with alanine (inactive in silica binding) in either position 6 or 11 of the peptide, the importance of its location was studied. The results showed that histidine at 11th position was important, as replacing that resulted in loss of binding. On the other hand, replacing histidine at 6th position had no detrimental effect, in fact, the binding was improved due to increased conformational flexibility, leading to additional hydrogen bonding (figure 9.10, bottom). These studies provided the following key mechanistic information:

- Different silica sources/synthesis methods exhibit differences in surface properties (ranging from subtle to dramatic).
- Peptides recognise these differences and therefore resultant peptide sequences are different.
- Surface ionisation is an important factor affecting peptide binding—changes in this ionisation can occur from differences in particle sizes or slight changes to synthesis condition or the solvent conditions (pH, ionic strength, and type of buffer).
- The peptide–silica interactions predominantly involved a combination of ion pairing and hydrogen bonds.
- Peptide conformation and its rigidity/flexibility were important. Changes in certain amino acids or at specific location in the sequence led to dramatic changes in binding, even leading to the identification of stronger binders that biopanning did not provide.

9.4.4 Do we need peptides or biomolecules?

This is a question that has been debated widely, not just in the case of silica but also for other minerals. One school of thought is that proteins and peptides are required because they provide molecular recognition, specificity, and selectivity. This is perhaps true *in vivo*, however, *in vitro*, it is worth verifying. As such, the other school of thought is that if the chemical and physical functionalities can be mimicked in synthetic molecules, then there may not be a need for specific peptides or proteins. It is important to recognise here that the ultimate goal is not to mimic biological *structures*, but to mimic the biological *process* in order to develop a bioinspired process. Hence, the need for the peptides/biomolecules may not be so important if synthetic molecules can mimic their function and thus the process. This is supported by some of the examples shown above (the use of polylysine and functionalised gold nanoparticles). Further, it is worth mentioning that bioextracts have been shown to facilitate the formation of non-specific materials. For example, silicatein and silaffin both can catalyse the formation of titania and other materials [68]. Particularly for non-crystalline materials, epitaxial matching between the biomolecules and the materials is of less relevance, further suggesting the possibility of successfully using synthetic molecules.

In order to address this point and other questions listed in section 9.4, a range of synthetic polymers and block co-polymers, small amines and dendrimers have been investigated. The chemistry of these systems was selected based on amine functionalities present either in the backbone (mimicking peptides/proteins) or in the sidechains (mimicking propylamines), see figure 9.6. The advantage of using synthetic molecules is that they are typically well-characterised and their properties readily measured.

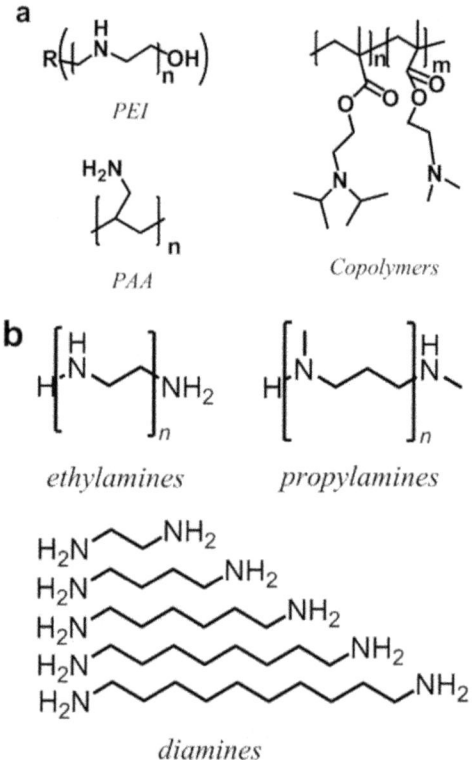

Figure 9.11. Selected examples of the bioinspired additives used in silica formation *in vitro*: (a) synthetic polymers and (b) small molecules [11], reproduced by permission of The Royal Society of Chemistry.

Examples of synthetic amine-containing polymers and block co-polymers include polyethyleneimines (PEI, linear, and branched), polyallylamine (PAA), and various versions of poly(alkylamino methacrylate) copolymers (see figure 9.11(a)). For a wider range of polymeric additives used, see [11]. Each of these polymers were found to catalyse the formation of silica, within seconds to minutes, under ambient conditions. Further, given their chain lengths, these molecules have the propensity to self-assemble into interesting template shapes. For example, in response to the processing conditions (e.g. pH and solvents), a star-shaped PEI assembles into a range of structures. Exploiting this dual catalytic and self-assembling function of many polymers, it is possible to produce a variety of silica morphologies including particles, plates, nanofibers, and coatings (see figure 9.12). This and other research clearly highlights the fact that there is no need to use biological molecules (peptides, proteins, or enzymes) to control silica formation. This is further supported by a large body of literature on the use of synthetic polymers as catalysts and templates for the bioinspired synthesis of wide ranging morphologies of silica (many of which are documented in [11]).

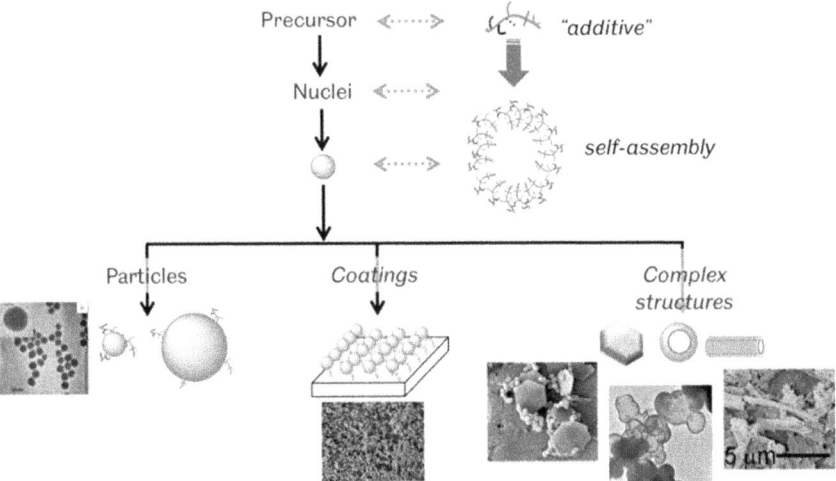

Figure 9.12. Schematic showing precursor to final bioinspired silica structures formed in the presence of additives. The scheme highlights the importance of key chemical moieties of the additives as well as their ability to self-assemble and to interact with a range of silicate species. By controlling these interactions, a range of materials can be produced. Image adapted from [72], copyright 2014 with permission of Springer.

9.4.5 Can smaller molecules provide similar activities?

We have seen that one group of the bioextracts isolated from diatoms is a family of propylamines with variable molecular size (or chain length) and levels of methylation. These propylamines have shown great activity in silica formation, both on their own as well as when mixed with silaffins. These propylamines are much smaller than siliaffins, silicateins and other biomolecules identified from biosilica. This implies that in addition to polymeric additives, smaller additives are likely to be functional. Obviously, the mechanisms of their action are of great interest, and scientists have been curious about the length of these amines, the level of amine methylation, and the size of the carbon spacers between amines. These points were systematically addressed by designing additives, for example those shown in figure 9.11(b) [57, 69–71]. The following key principles emerged in the context of bioinspired silica synthesis [57, 69–71]:

- Additives need to be positively charged throughout silica formation process.
- Additives need to be partially charged under the synthesis conditions, such that they can reversibly undergo protonation and deprotonation.
- When only the alternate amines are charged, the additives seem to provide the best results in terms of silica formation and precipitation.
- The balance between the hydrophilic and hydrophobic parts of the additives controls the catalysis and templating.
- For a given chain length (or number of amine groups), the 3-carbon spacing between the amine groups (as seen in biosilica-associated propylamines) was the best in achieving highest catalysis, compared with 2- or 4-carbon spacing (seen in ethylamines or butylamines, respectively).

- Additives with longer chain lengths (i.e. more amines per molecule) performed better, however, molecules as small as 3–5 amines groups were sufficient.

9.5 Benefits of bioinspired synthesis

The understanding of molecular interactions between additives discussed above and silica has now allowed the rapid synthesis of silica under mild conditions (see figure 9.13). It is worth noting that although organic molecules were used as templates in the formation of silicas and silicates at least a decade before the advent of bioinspired silica (e.g. zeolites and mesoporous silicas), what is new is the bioinspired catalytic ability of these additives. This dual function has enabled the synthesis of silica of desired shapes/morphologies which occurs:
- at room-temperature,
- at neutral pH,
- in water (without the need for co-solvents),
- at ambient pressure,
- very fast (within seconds to minutes), and
- as a single step synthesis.

Further, this method produces comparatively little waste, yet provides superior control of product properties. In contrast, for example, traditional syntheses of mesoporous silica suffer seriously from a range of issues, such as the use of toxic precursors, long synthesis (2–6 days), requirement of hydrothermal conditions (elevated temperature and pressure), and extremes of pH. This is a profound step change in the synthesis of silica-based materials—we now have a green/sustainable method for producing high value silicas. These differences between bioinspired and traditional syntheses of a range of silicas (including commercial products) are summarised in table 9.1. The details of the synthesis procedure are documented in [73], and also in the following videos:
- *Preparation of functional silica using a bioinspired method*: https://www.jove.com/video/57730/preparation-of-functional-silica-using-a-bioinspired-method.
- Small scale synthesis podcast: https://youtu.be/dkwsnrGesbA.

In addition to understanding the *end-product* (i.e. the properties of the silica synthesised), it is also important to understand the *process* by which these materials were formed. Therefore, probing the particle formation pathways and the roles of the additives in this process is of importance. As depicted in figure 9.14, the silica formation process is complex and undergoes multiple stages. During these stages,

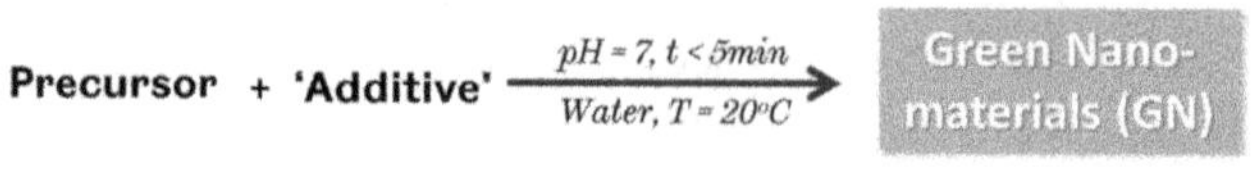

Precursor + 'Additive' $\xrightarrow[\text{Water, } T = 20°C]{pH = 7, t < 5min}$ Green Nanomaterials (GN)

Figure 9.13. Simplified overview of bioinspired synthesis of silica.

Table 9.1. Comparison of selected traditional methods with bioinspired synthesis of silica, adapted from [75].

Silica type	Reagents[a]	Solvents	Reaction conditions				Finishing	Control over
			t h	T °C	pH			
Mesoporous—SBA-15	TEOS, pluronic	Water, ethanol	44	42–100	<7		Calcination at 550 °C (>6 h)	Pore size, space group, wall thickness
Industrial—precipitated	Silicate, H_2SO_4	Water	2.5–3	60–80	<7		Drying	Purity, dispersion
Industrial—gel	Silicate, H_2SO_4	Water	3–5	35–80	ca. 7		Sizing, washing, drying	Pore volume
Colloidal—Stöber	TEOS and ammonia	Water, ethanol	12–24	10–60	ca. 9		Centrifugation, drying	Pore size
Bioinspired	Silicate and additive	Water	0.08	20	7		Centrifugation, drying	Pore size, structure, porosity, particle size, morphology,.....[b]

[a] Silicate = sodium silicate or water glass, CTAB = cetyl trimethylammonium bromide, TEOS = tetraethoxysilane, pluronic = a triblock co-polymer of ethylene- and propylene-oxides.
[b] Control over incorporation of foreign material (catalyst, enzymes, drugs, etc) is also possible.

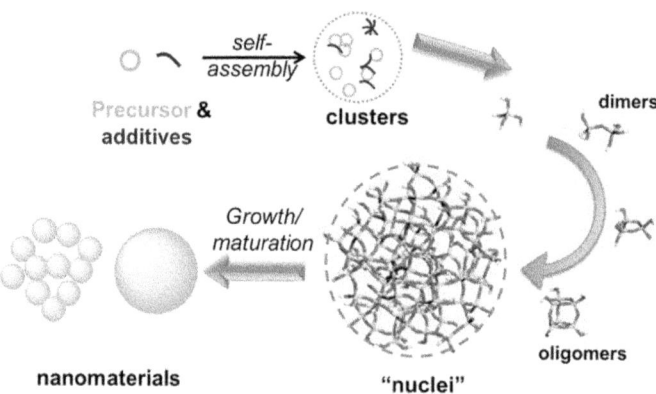

Figure 9.14. Stages in silica formation pathways, showing some of the key molecules/species involved. Adapted from [74].

the concentration of various silicon species (monomer and oligomers in particular) evolve. The additives interact in each stage and play a role in determining the formation pathways taken. Further depth of knowledge created through understanding the formation process in the presence of additives can make it possible to control the synthesis, and hence produce a range of commercially important properties of silica (e.g. porosity, particle sizes, and surface chemistry).

As we have seen in this chapter and chapter 6, biology offers sophistication. In the case of silica, although biosilica from plants and sponges is less ornate, diatoms show a remarkable nanoscale and hierarchical assembly. This spans nearly four orders of magnitudes and ranges from the primary particles (5–10 nm) all the way to the size of the cell (which varies for species between 10 μm to 100 μm). This assembly offers a multiscale structure made up of particles and pores. As this is not just controlled by the biomolecules but also involves other components of cellular machinery and is under strict biological control, such hierarchical assembly is incredibly difficult to obtain on the bench top using bioinspired methods, much less at industrial scales. Better understanding the particle formation process as shown in figure 9.14 is likely to help make advances in obtaining sophisticated features observed in biosilica. Indeed, we have already seen some complex structures being synthesised using bioinspired approaches, such as those shown in figure 9.9; note that biology is not known to produce these kinds of structures. On the other hand, the aim of bioinspired synthesis is not to copy the biomineral structures but to copy the principles and utilise them for producing desirable products. As diatomite or diatomaceous earth, which is essentially fossilised diatoms, is readily available in large quantities, the question to consider is whether we need to produce synthetic 'diatom-like' structures.

9.6 From lab to market

So, using this knowledge, what are we able to do, and what should we do? The ability to control particle formation pathways and the attributes of silica product using a green method is a great and novel outcome. If the goal is to develop

industrially viable methods for commercial silicas, then the following attributes are important:
- Purity of the silica produced. This is a really important property, which is generally not given due importance during lab-scale studies. In particular, as the additives (and salt by-products) get occluded in final silicas, the purity of silica should be checked.
- Porosity is the other key property for silica. In most applications, the surface area, pore sizes, and pore volumes govern their performance. These applications include desiccant, adsorption platforms for pollutant removal from air, water and oils, fillers for polymers, cosmetics and toothpastes, and excipients for pharmaceuticals.
- Yield relates to productivity and waste generated. Although the yield is rarely reported in the synthesis literature, the economic viability of the method and the materials largely depend on it.
- Morphology is useful to control, however, in commercial applications, it is not so critical. In contrast, scientific publications put a lot of emphasis on morphology as it can provide information on growth mechanisms.
- Particle size, sphericity, and polydispersity are more related to the handling of solid particles, however, they may also contribute towards the material performance.
- pH and surface silanols provide a good indication of silica compatibility with various formulations (e.g. cosmetics or tyres).

Above, we have only focused on the 'product' attributes. In addition, there are 'process' considerations when it comes to industrial production, such as unit operations and scale-up. Given the focus of this book on materials, the process features are not discussed here. Nevertheless, this feat in controlling bioinspired silica has resulted in a number of patent applications and commercialisation activities. Some of these are listed below, in chronological order of their first disclosure:
- WO2000035993: *Methods, compositions, and biomimetic catalysts for in vitro synthesis of silica, polysilsesquioxane, polysiloxane, and polymetallo-oxanes.* This invention came from the University of California (Santa Barbra), led by Morse. This invention focuses on the use of silicatein and its mimics (polypeptides) in catalysing the formation of a range of silicon-containing materials under 'green' conditions.
- US7169589B2: *Silicatein-mediated synthesis of amorphous silicates and siloxanes and use thereof.* Müller and Schröder (Johannes Gutenberg University, Mainz) disclosed the use of silicatein (natural or recombinant) in the synthesis of silica.
- US7335717B2: *Methods, compositions, and biomimetic catalysts for the synthesis of silica, polysilsesquioxanes, polysiloxanes, non-silicon metalloid-oxygen networks, polymetallo-oxanes, and their organic or hydrido conjugates and derivatives.* This invention also came from the University of California

(Santa Barbra), led by Morse. Similar to their earlier invention, this patent widens to the use of synthetic additives.
- US2003171525: *Structure-directing catalysis for synthesis of metal, non-silicon metalloid and rare earth oxides and nitrides, and their organic or hydrido conjugates and derivatives.* Following from their earlier work, Morse and co-workers extended the use of silicatein (and its mimics) in catalysing the formation of a range of non-silicate materials.
- US7960509B2: *Fibrous protein fusions and use thereof in the formation of advanced organic/inorganic composite materials.* Kaplan (Tufts University) and Naik (Air Force Research Lab) designed a range of fusion proteins that combine silk with silica forming peptides and disclosed their use in the synthesis of hybrid materials, including silica.
- US8383755B2: *Enzyme-medicated cross-linking of silicone polymers.* Zelisko and co-workers (Brock University) developed the use of a range of enzymes, inspired from silicatein and silaffin, for catalysing siloxane bond formation.
- US2008293096: *Enzyme and template-controlled synthesis of silica from non-organic silicon compounds as well as aminosilanes and silazanes and use thereof.* Another invention from the group of Müller and Schröder, which extends their previous invention into the use of a range of proteins, enzymes and peptides for the synthesis of silicates, non-silicates and silanes.
- WO2008022774: *Biosilica-adhesive protein nanocomposite materials: synthesis and application in dentistry.* This invention from the group of Müller and Schröder is on the design of silicatein-silk fibroin fusion proteins and their use in synthesising silica-based materials for dental filling applications.
- WO2010036344: *Compositions, oral care products and methods of making and using the same.* This invention came from the group of Müller and Schröder from Johannes Gutenberg University (Mainz), in collaboration with Grace (a silica manufacturer). It pertains to the direct use of silicatein to deposit protective inorganic coatings on teeth.
- US2011281077: *Silicon derivate layers/films produced by silicatein-mediated templating and process for making the same.* This invention arising from a collaboration between the groups of Dario Pisignano (Pisa University) and Müller and Schröder, developed a method for producing silica coatings using silicatein.
- WO2015150399: *Osteogenic material to be used for treatment of bone defects.* Müller and Schröder further extended the use of silicateins for bone regeneration applications.
- WO2017037460: *Silica synthesis.* This invention, from one of the authors of this book, disclosed a bioinspired green method to synthesise silica and to further purify it by removing the occluded additives at room temperature and in water.

One of the authors of this book has embarked on the journey from lab-based green synthesis to large-scale manufacturing. The highlights of this journey are summarised below, in order to provide a better understanding of the pathway taken

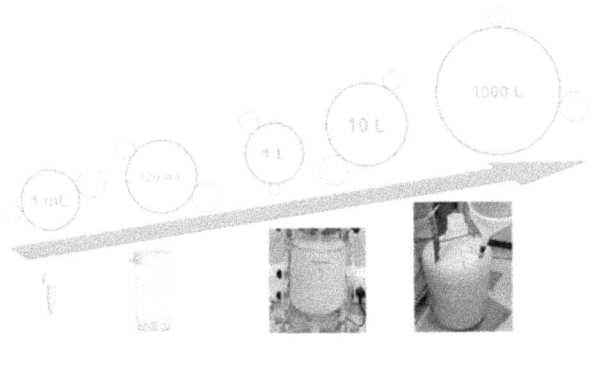

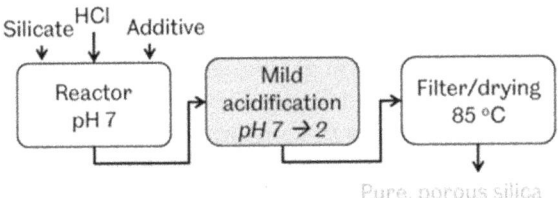

Figure 9.15. Pictorial representation of our scale-up journey from 2 ml scale to 40 L scale (top). The new purification method developed is shown as a block flow diagram (bottom). Images reproduced from [76].

[76]. A systematic approach was taken for process scale-up and process intensification. Initially a techno-economic analysis of a bioinspired method was performed in order to assess the economic feasibility [77]. The results show that using our green methods it is possible to reduce the energy usage of the reaction step by ~95% when compared with a traditional process, and the green nanomaterials (GN) would cost the same as the lowest grade commercial counterparts, yet provide significantly better quality and properties. This is promising and supports further development work (see figure 9.15, top, for a schematic showing the scale-up journey).

It is also worth focusing on aspects of downstream processing, in particular, purification of the products. In order to render porosity and purity for organic-mediated synthesis, calcination is used. However, it is prohibitively energy intensive and strips off any green benefits claimed from the synthesis step. Solvent reflux forms an alternative, but is also energy intensive, as it requires solvent reflux. This means that all the green features of bioinspired synthesis are outweighed by the purification step. Although downstream separations and purifications are generally more energy intensive/expensive, this creates economic barriers to commercialisation and scale-up—the bioinspired method thus cannot add value over existing processes and products. In order to address these issues, a new method for purification was invented, which involves mild acidification, operates at room temperature, and takes a few minutes (figure 9.15, bottom, and WO2017037460A1) [78]. This new method allows a complete removal of organics, with an added possibility of composition and porosity control. Given that it is a non-destructive method, most of the water and additive can be recycled, further improving the sustainability and economics. It is worth noting here that in terms of feedstock, the additives cost significantly more than the silicate precursors. Therefore, the ability to recover and reuse additives is of immense importance—this is not possible with destructive methods such as calcination.

Further, a number of scalability assessments of this method were performed (see figure 9.15, top, and an associated video of reaction in a 5 L reactor: https://youtu.be/S5YvvontRVA). Initially, the synthesis was carried out in lab-scale continuous flow reactors (at small scale: 5–50 ml) [79]. These studies provided information on key process parameters, as well as identifying practical challenges. Learning from these results and refining the process, the scale could be increased from a few mL to 1 L, 5 L, and reaching 40 L. The process was tested in both batch and continuous mode, and in tank and tubular reactors—this is an important step in assessing the scalability of a new method. Key learning from these scale-up trials included that the method is readily scalable and the recoverable yield does not change with scale. Further, as expected, mixing significantly affected the material properties [80]. These findings are currently being utilised in commercialising the process as well as up-scaling to produce kilograms of products.

9.7 Summary: key learning, summary and the future

We have learnt in this chapter that intricate biosilica structures are deposited by a range of living systems, and that they do this via an extremely complex process. Molecular biologists and biochemists have studied these systems extensively over decades in order to reveal the molecular secrets of biosilicification. They have been able to isolate genes, cellular components and biomolecules that control anything from silicon uptake to biosilica deposition. Further, *in vitro* studies of these biomolecules have started to provide key information that can be used to develop green synthesis protocols. These outcomes have spurred on interest in developing synthetic additives that can mimic the function of biomolecules in order to synthesise silica in a controlled and a sustainable fashion (see figure 9.16 also see a summary podcast at https://youtu.be/sDUl7urlsxY). In this journey of bioinspired silica, numerous intriguing features of silica formation and silicate-additive interactions have been unveiled. Some of these new concepts include:
- Cationic molecules that are readily water soluble are generally useful in facilitating silica formation under ambient conditions and neutral pH.
- Additives interact with different (and perhaps selective) stages of silica formation, which leads to the differences in their actions and the features of silica produced.
- Dynamic/reversible protonation of the additives is important.
- Additives can self-assemble or co-assemble with silicates, leading to templating final structures.

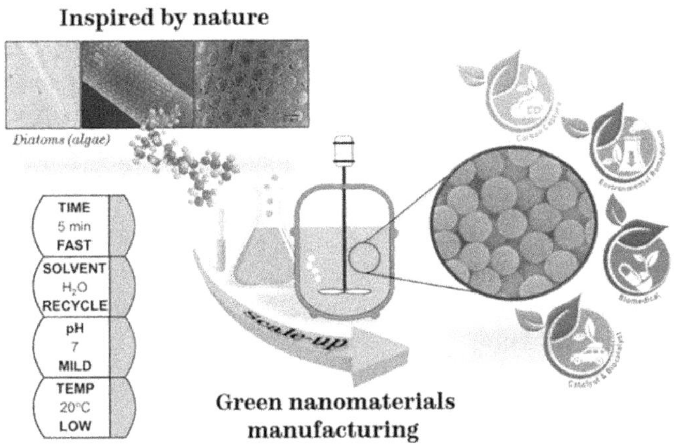

Figure 9.16. An overview of the journey from biosilica to the invention of bioinspired silica, the green credential, scale-up and applications. Image reproduced from [75], copyright 2015 with permission from Elsevier.

- The structure, architecture and amine environment, and the length of the additive play crucial roles in controlling silica synthesis and material properties.

The future focus should be on developing robust science underpinning the correlations between the synthesis–structure–property–performance for these materials, so that they can be easily applied to existing and emerging markets. Scientists should be working with industry to develop these materials for specific applications, and collaborating with engineers to design new sustainable/green manufacturing methods.

References

[1] Femmer R 2016 Diatoms (50 Species) (https://www.usgs.gov/media/images/diatoms-50-species-3)
[2] Pisera A 2003 *Microscop. Res. Tech.* **62** 312
[3] Brook M A 2000 *Silicon in Organic, Organometallic, and Polymer Chemistry* (New York: Wiley)
[4] Simpson T L and Volcani B E 1981 *Silicon and Siliceous Structures in Biological Systems* (New York: Springer)
[5] Evered D and O'Connor M 1986 *Ciba Foundation Symp.* vol 121 (New York: Wiley)
[6] Perry C C and Keeling-Tucker T 2000 *J. Biol. Inorg. Chem.* **5** 537
[7] Hildebrand M 2000 *Biomineralization: From Biology to Biotechnology and Medical Application* ed E Baeuerlein (Weinheim: Wiley), p 171
[8] Hildebrand M 2008 *Chem. Rev.* **108** 4855
[9] Hildebrand M and Wetherbee R 2003 *Prog. Mol. Subcell. Biol.* **33** 11
[10] Perry C C 2009 *Biosilica in Evolution, Morphogenesis, and Nanobiotechnology: Case Study Lake Baikal* vol 47 ed W E G Müller and M A Grachev (Berlin: Springer), p 295
[11] Patwardhan S V 2011 *Chem. Commun.* **47** 7567

[12] Harrison C C 1996 *Phytochemistry* **41** 37
[13] Kröger N, Lehmann G, Rachel R and Sumper M 1997 *Eur. J. Biochem.* **250** 99
[14] Hildebrand M, Volcani B E, Gassmann W and Schroeder J I 1997 *Nature* **385** 688
[15] Shimizu K, Cha J, Stucky G D and Morse D E 1998 *Proc. Natl. Acad. Sci. U.S.A.* **95** 6234
[16] Krasko A, Lorenz B, Batel R, Schröder H C, Müller I M and Müller W E G 2000 *Eur. J. Biochem.* **267** 4878
[17] Perry C C and Keeling-Tucker T 2003 *Colloid Polym. Sci.* **281** 652
[18] Ma J F, Tamai K, Yamaji N, Mitani N, Konishi S, Katsuhara M, Ishiguro M, Murata Y and Yano M 2006 *Nature* **440** 688
[19] Frigeri L G, Radabaugh T R, Haynes P A and Hildebrand M 2006 *Mol. Cell. Proteomics* **5** 182
[20] Matsunaga S, Sakai R, Jimbo M and Kamiya H 2007 *Chembiochem.* **8** 1729
[21] Kroger N and Poulsen N 2008 *Annu. Rev. Genet.* **42** 83
[22] Ehrlich H *et al* 2010 *Nat. Chem.* **2** 1084
[23] Patwardhan S V, Clarson S J and Perry C C 2005 *Chem. Commun.* 1113
[24] Armbrust E V *et al* 2004 *Science* **306** 79
[25] Bowler C *et al* 2008 *Nature* **456** 239
[26] Kröger N, Deutzmann R, Bergsdorf C and Sumper M 2000 *Proc. Natl. Acad. Sci. U.S.A.* **97** 14133
[27] Kröger N, Deutzmann R and Sumper M 1999 *Science* **286** 1129
[28] Kröger N, Deutzmann R and Sumper M 2001 *J. Biol. Chem.* **276** 26066
[29] Kröger N, Lorenz S, Brunner E and Sumper M 2002 *Science* **298** 584
[30] Poulsen N and Kröger N 2004 *J. Biol. Chem.* **279** 42993
[31] Poulsen N, Sumper M and Kröger N 2003 *Proc. Natl. Acad. Sci. U.S.A.* **100** 12075
[32] Scheffel A, Poulsen N, Shian S and Kröger N 2011 *Proc. Natl. Acad. Sci.* **108** 3175
[33] Cha J N, Shimizu K, Zhou Y, Christiansen S C, Chmelka B F, Stucky G D and Morse D E 1999 *Proc. Natl. Acad. Sci. U.S.A.* **96** 361
[34] Zhou Y, Shimizu K, Cha J N, Stucky G D and Morse D E 1999 *Angew. Chem. Int. Ed.* **38** 780
[35] Schröder Heinz C, Vlad A G, Xiaohong W and Werner E G M 2016 *Bioinspir. Biomim.* **11** 041002
[36] Perry C C and Keeling-Tucker T 1998 *Chem. Commun.* **23** 2587
[37] Currie H A and Perry C C 2007 *Ann. Bot.* **100** 1383
[38] Currie H A and Perry C C 2009 *Phytochemistry* **70** 2089
[39] Gong N, Wiens M, Schroder H C, Mugnaioli E, Kolb U and Muller W E G 2010 *J. Exp. Biol.* **213** 3575
[40] Hildebrand M, Lerch S J L and Shrestha R P 2018 *Front. Mater. Sci.* **5** 125
[41] Brutchey R L and Morse D E 2008 *Chem. Rev.* **108** 4915
[42] Schröder H C, Grebenjuk V A, Wang X and Müller W E G 2016 *Bioinspir. Biomim.* **11** 041002
[43] Murr M M and Morse D E 2005 *Proc. Natl. Acad. Sci. U.S.A.* **102** 11657
[44] Brunner E, Lutz K and Sumper M 2004 *Phys. Chem. Chem. Phys.* **6** 854
[45] Sumper M 2002 *Science* **295** 2430
[46] Patwardhan S V, Holt S A, Kelly S M, Kreiner M, Perry C C and van der Walle C F 2010 *Biomacromolecules* **11** 3126
[47] Groeger C, Lutz K and Brunner E 2008 *Cell Biochem. Biophys.* **50** 23

[48] Knecht M R and Wright D W 2003 *Chem. Commun.* **24** 3038
[49] Mizutani T, Nagase H, Fujiwara N and Ogoshi H 1998 *Bull. Chem. Soc. Jpn.* **71** 2017
[50] Cha J N, Stucky G D, Morse D E and Deming T J 2000 *Nature* **403** 289
[51] Brott L L, Naik R R, Pikas D J, Kirkpatrick S M, Tomlin D W, Whitlock P W, Clarson S J and Stone M O 2001 *Nature* **413** 291
[52] Coradin T and Livage J 2001 *Colloids Surf. B-Biointerfaces* **21** 329
[53] Patwardhan S V, Mukherjee N and Clarson S J 2001 *J. Inorg. Organomet. Polym.* **11** 193
[54] Sudheendra L and Raju A R 2002 *Mater. Res. Bull.* **37** 151
[55] Patwardhan S V and Clarson S J 2003 *J. Inorg. Organomet. Polym.* **13** 49
[56] Patwardhan S V and Clarson S J 2003 *J. Inorg. Organomet. Polym.* **13** 193
[57] Belton D, Paine G, Patwardhan S V and Perry C C 2004 *J. Mater. Chem.* **14** 2231
[58] Kisailus D, Najarian M, Weaver J C and Morse D E 2005 *Adv. Mater.* **17** 1234
[59] Patwardhan S V, Mukherjee N, Steinitz-Kannan M and Clarson S J 2003 *Chem. Commun.* **10** 1122
[60] Patwardhan S V, Maheshwari R, Mukherjee N, Kiick K L and Clarson S J 2006 *Biomacromolecules* **7** 491
[61] Cui H, Krikorian V, Thompson J, Nowak A P, Deming T J and Pochan D J 2005 *Macromolecules* **38** 7371
[62] Tomczak M M, Glawe D D, Drummy L F, Lawrence C G, Stone M O, Perry C C, Pochan D J, Deming T J and Naik R R 2005 *J. Am. Chem. Soc.* **127** 12577
[63] Bellomo E G and Deming T J 2006 *J. Am. Chem. Soc.* **128** 2276
[64] Patwardhan S V, Emami F S, Berry R J, Jones S E, Naik R R, Deschaume O, Heinz H and Perry C C 2012 *J. Am. Chem. Soc.* **134** 6244
[65] Emami F S, Puddu V, Berry R J, Varshney V, Patwardhan S V, Perry C C and Heinz H 2014 *Chem. Mater.* **26** 5725
[66] Emami F S, Puddu V, Berry R J, Varshney V, Patwardhan S V, Perry C C and Heinz H 2014 *Chem. Mater.* **26** 2647
[67] Sola-Rabada A, Michaelis M, Oliver D J, Roe M J, Ciacchi L C, Heinz H and Perry C C 2018 *Langmuir* **34** 8255
[68] Dickerson M B, Sandhage K H and Naik R R 2008 *Chem. Rev.* **108** 4935
[69] Belton D, Patwardhan S V and Perry C C 2005 *Chem. Commun.* **27** 3475
[70] Belton D J, Patwardhan S V, Annenkov V V, Danilovtseva E N and Perry C C 2008 *Proc. Natl. Acad. Sci. U.S.A.* **105** 5963
[71] Belton D J, Patwardhan S V and Perry C C 2005 *J. Mater. Chem.* **15** 4629
[72] Patwardhan S 2014 *Bio-Inspired Nanotechnology: From Surface Analysis to Applications* ed M R Knecht and T R Walsh (New York: Springer), 127
[73] Manning J R H, Routoula E and Patwardhan S V 2018 *J. Vis. Exp.* **138** e57730
[74] Belton D J, Deschaume O, Patwardhan S V and Perry C C 2010 *J. Phys. Chem. B* **114** 9947
[75] Patwardhan S V, Manning J R H and Chiacchia M 2018 *Curr. Opin. Green Sustain. Chem.* **12** 110
[76] Patwardhan S V 2019 *Johnson Matthey Technol. Rev.* **63** 152
[77] Drummond C, McCann R and Patwardhan S V 2014 *Chem. Eng. J.* **244** 483
[78] Manning J R H, Yip T W S, Centi A, Jorge M and Patwardhan S V 2017 *Chem. Sus. Chem.* **10** 1683
[79] Patwardhan S V and Perry C C 2010 *Silicon* **2** 33
[80] Chiacchia M and Patwardhan S V 2018 *16th European Conf. on Mixing (Toulouse, France)*

Lightning Source UK Ltd.
Milton Keynes UK
UKHW032006310120
357972UK00005B/198